D0124436

WHEN A LOOSE CANNON FLOGS A DEAD HORSE THERE'S THE DEVIL TO PAY

Olivia A. Isil

Seafaring Words in Everyday Speech

INTERNATIONAL MARINE
CAMDEN, MAINE

International Marine/
Ragged Mountain Press

A Division of The McGraw-Hill Companies

10   9   8

Copyright © 1996 International Marine®, a division of The McGraw-Hill Companies.
All rights reserved. The publisher takes no responsibility for the use of any of the materials or
methods described in this book, nor for the products thereof. The name "International Marine"
and the International Marine logo are trademarks of The McGraw-Hill Companies. Printed in
the United States of America.

Library of Congress Cataloging-in-Publication Data
Isil, Olivia A.
   When a loose cannon flogs a dead horse there's the devil to pay :
seafaring words in everyday speech / Olivia A. Isil.
      p. cm.
   Includes bibliographical references and index.
   ISBN 0-07-032877-3
   1. English language—Terms and phrases. 2. Naval art and science—Terminology.
3. English language—Etymology. 4. Seafaring life—Terminology. 5. Sailors—Language.
I. Title.
PE1583.I85   1996
422'.0883875—dc20   96-4236
                                                  CIP

Questions regarding the content of this book should be addressed to:

International Marine, P.O. Box 220, Camden, ME 04843, 207-236-4837

Questions regarding the ordering of this book should be addressed to:

The McGraw-Hill Companies, Customer Service Department, P.O. Box 547,
Blacklick, OH 43004; retail customers: 1-800-262-4729; bookstores: 1-800-722-4726

This book is printed on 60-pound Renew Opaque Vellum, an acid-free paper that contains
50 percent recycled waste paper (preconsumer) and 10 percent postconsumer waste paper. ♻

Typeset in 11-point Adobe Garamond
Printed by R. R. Donnelley, Crawfordsville, IN
Design by Carol Gillette
Production and page layout by Janet Robbins
Edited by John J. Kettlewell, Nancy C. Hauswald, and Pamela Benner

Permission to reprint selections from "Sailor Man," by H. Sewell Bailey, and "To the Humpback
Whales," by Harold J. Morowitz, which appeared in *Rhyming in the Rigging: Poems of the Sea*
(© 1978, Ox Bow Press), was granted by the publisher.

CONTENTS

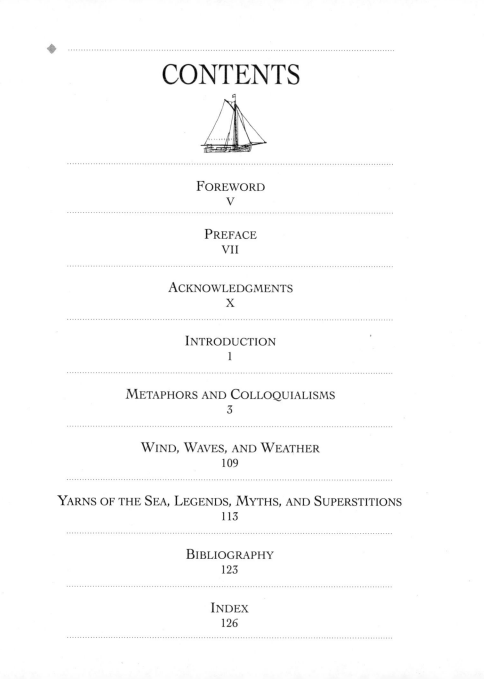

FOREWORD

Peter Stanford, President, National Maritime Historical Society

The sea, to Olivia Isil, is more than a place. "It is a state of mind," she says, "a condition of the soul." Isil knows that there is something special, too, about sailors, and about the way they live, act, and speak. She has combined her long-standing fascination with the magic of the sea and her professional skills as a historian and wordsmith to write this remarkable compendium of sea language that has come ashore.

These seagoing words and phrases that have found a secure berth in our everyday speech were not invented for your reading pleasure. "Keep a weather eye open," for example, means just what it says. At sea, trouble generally comes from the direction of the wind—the "weather" side—and a ship that does not keep an alert lookout for what's coming next from that direction is asking for trouble, and possibly a one-way trip to Davy Jones's locker.

And ashore? Seaport taverns are grand places for exchanging ideas and information. There's no doubt in my mind, at least, that that's where most sea language came ashore. Once a landlubber has heard an expression like "keep a weather eye open," who is going to be content with a tame "be careful," or even "watch out for trouble"?

Of course, life is not an endless blue Monday for sailors, chock-full of trouble, and they're not always between the devil and the deep blue sea. Like the rest of us, they sometimes find things running on an even keel, with plain sailing ahead, and everything shipshape and Bristol fashion. You'll find these phrases, and many more, traced to their actual origins in seafaring practice in this book.

The author also chronicles sailors' lore about changing weather conditions and takes real joy in exploring sailors' myths and traditions, from the practice of christening a ship when she is first launched, to Fiddler's Green, the seaport town where sailors go when they cut their painters.

Historically, sailors are a conservative lot, skeptical of newfangled notions and ideas. Their commitment to tradition has enriched our lives; they have preserved the colorful language of the sea, born in a harsh, unforgiving, and challenging environment, for our use and enjoyment ashore.

In the ensuing pages is a celebration of the richly creative flowering in words and thought that conservative, action-oriented sailors have given to all hands who speak the English language.

PREFACE

hy does the sea hold such a fascination for mere mortals? Does it cast a magic spell? Do we project our own doubts into the turbulent waves and rush of the tides and find release of our tensions and fears in the calm that follows the crush of noisy waves? Are we fascinated by, or afraid of, its paradoxes, its awesome beauty, its destructive power, its shifting moods? The lure of the sea is timeless:

> *. . . Yet still, even more now, my spirit within me*
> *Drives me seaward to sail the deep,*
> *To ride the long swell of the salt sea waves.*
> *Never a day but my heart's desire*
> *Would launch me forth on the long sea path . . .*
> —"The Seafarer"

Written more than eleven centuries ago by an unknown sailor, "The Seafarer" is one of the oldest poems extant in English literature. Neither rustic boats nor tall ships play a significant role in modern life, but the sea has lost none of its fascination. Like the ancient Seafarer, most of us still crave adventure, still long to test our mettle against the odds, still value individual accomplishment, still long for the kind of peace that comes with self-worth, still are lured by the sea.

My own fascination with the sea began in childhood, when I played on the rotting decks of the U.S. *Niagara.* She had never seen blue water, nor was I to until many years later, but the sight of the *Niagara* cradled on the shore with the expanse of Lake Erie as a background captured my imagination. In my mind's eye, the lake became the Spanish Main, the Cape of Good Hope, and the "roaring forties," and the decks of the old ship bristled with explorers, Elizabethan sea dogs, American patriots, and daring characters who sprang to life from the pages of Melville, Smollett, and other storytellers to be my shipmates. Queequeg, Jack Nastyface, and Peregrine Pickle joined the stage of my playground/library/learning lab and enriched my life with an avid interest in history and the sea. Their creators imbued me with a love of language and an interest in a well-turned phrase.

My passion for words and their origins was further enhanced by my maternal grandmother, whose first language was not English. Her spectacular malapropisms, "danglers," and mispronunciations were amusing and colorful. Her tapestry of language was so vibrant that her stories and the people she described remained alive for her listeners for many years. My dad, a diver-dockworker during World War II, and my three sailor uncles also influenced my interest in the lively language of the sea. With a watch cap jammed down around my ears, bos'n's pipes dangling from my neck, I was always perched, parrotlike, on one of their shoulders. Much to my father and uncles' sheer delight, and my mother's extreme horror, I was a perfect mimic and repeated everything (and I do mean everything!) that came within earshot. It was easy enough for Mother to track down the source of my vivid language. Dad spoke English with a slight accent and liberally enriched his sentences with a zinger or two in Italian. At the age of six, I was packed off to the Sisters of St. Joseph and the refined environment of a convent school—probably just in the nick of time.

Under the gentle discipline and constant tutelage of the sisters, the handiwork of my former language mentors was soon undone, at least when I spoke. I never really forgot, however, nor have I ever been cured of, any of that wonderful childhood case of sea fever. In my academic and religious studies, I found the imagery of the sea in religious metaphors and allegories: "I wander as in a fragile bark on life's tempestuous seas" . . . "anchors of hope" . . . "harbors of salvation." Daily history lessons and the fate of nations were played out at sea; my beloved *Niagara,* outgunned and outmanned, sailed directly into a British squadron with her carronades blazing. As a student of Latin, I sailed perilous voyages with Odysseus, with Jason and the Argonauts. If a math exam was in the offing, I could always count on a problem based on speed in nautical miles. The influence of the sea ebbed and flowed through art classes, in powerful visual impressions by Monet and Winslow Homer, and in music classes, in majestic tonal seascapes by Debussy and Mussorgsky.

But nowhere was the influence of the sea so mesmerizing as in literature classes. From Beowulf's ship, which sped on "over the breaking billows, with bellying sail, / And foaming beak, like a flying bird . . ." to Coleridge's Ancient Mariner, struggling for his soul on a phantasmagoric sea; to Emily Dickinson's sea-inspired intoxication and exultation of spirit; to Masefield's poignant longing for "the windy green unquiet seas"; I followed the path presented to me and discovered that the sea is more than a place. It is a state of mind, a condition of the soul.

Since my introduction to the sea, and my early studies, life has taken me to both coasts of the United States and to Europe. I have finally seen, and loved, blue water.

ACKNOWLEDGMENTS

Although sheer numbers make it impossible to acknowledge everyone individually, this book exists largely through the encouragement and generous support of others.

Special thanks must go to my brother, Fred Ciccozzi, of Erie, Pennsylvania, whose computer skills saved his "illiterate" sister from disgrace; to my very dear sister-in-law, Cheryl Ciccozzi, who was instrumental in the preparation of the first manuscript; to Lebame Houston, of Manteo, North Carolina; Thomas L. White, Jr., also of Manteo; and Marcia Kay Thompson, of New York City—friends of the highest order, who gave generously of their time and offered many practical suggestions along the way. Lastly, I wish to acknowledge and thank the staff of International Marine for their patience, guidance, and unfailing good humor during the publishing process.

Despite the painstaking research that has gone into the writing of this book, it has been a true labor of love. In that spirit, I wish to acknowledge with gratitude and deepest affection my earliest "language tutors": my parents, grandparents, and "sailor uncles." I would be remiss if I did not acknowledge the Sisters of St. Joseph, who took over where the others left off. My family and the good sisters remain a special part of me and the inspiration for this book.

INTRODUCTION

*S*ea imagery and nautical sayings ebb and flow throughout contemporary English speech. Words and expressions originally coined to identify or describe activities, people, or things associated with the sea have slowly washed ashore. As a result, many nautical terms and expressions have taken on a new life as colorful metaphors, now incorporated in the language of the land-based populace. Many of these colloquialisms are so entrenched in modern speech that few who use them are aware they are borrowing from sailors' jargon of yore.

When a Loose Cannon Flogs a Dead Horse There's the Devil to Pay traces some popular colloquialisms to their nautical origins. In many cases, derivation from a nautical root is unquestionably valid, the history and semantic development of the word or phrase being relatively easy to document. Some expressions have a basis in fact, but because their meanings have altered so much over the years, a direct line of development can only be suggested. A few salty terms have been embellished by good storytellers over the centuries. The text also includes several expressions that are strictly apocryphal, but part of a rich oral tradition even so.

Throughout the text, each entry is followed by the colloquial use of the word or expression, not the strict nautical meaning. The examples cited, for the most part, illustrate the entry's metaphorical usage. In the interest of clarity, spelling from early sources has been modernized. Also, whenever I quote someone in the book, I have tried to give both a first and a last name; however, there are several authors of older publications whose first names simply aren't known to us today. In those cases, I have cited their initials only.

Now, with bearings set, and navigational charts in hand, embark on a short sail or a long voyage through the sea of words that awaits. Conditions may sometimes be foggy but, by and large, smooth sailing is in the offing.

METAPHORS AND COLLOQUIALISMS

A1 ... (the best)

In the late seventeenth century, the insurance firm Lloyd's of London issued an A1
rating to merchant ships whose hull and gear were of the highest quality.
Over the years, *A1* came into general usage as a reference to excellence of
any kind. Even characters in Charles Dickens's *Pickwick Papers* use the
term: "'He must be a first-rater,' said Sam . . . 'A1,' replied Mr. Roker."

Above Board ... (honest dealing)

Any activity that is synonymous with fair play and honesty, or takes place in plain view,
is considered to be "above board." Although the origin of the expression is
obscure, some modern authors suggest that it stems from its opposite,
"below board." In the age of piracy, disreputable captains, in pursuit of

vulnerable merchant ships, attempted to conceal the strength of their ship's complement by hiding their crews below the boards (deck)—a practice synonymous with foul play. It is, however, more likely that the expression originated in gaming etiquette, which dictates that players keep their hands on the table where they can be seen. As the celebrated English clergyman and author John Earle noted in his work *Microcosm, Or a Piece of the World Discovered in Essays and Characters* (1628), "one that does it fair and above board [plays] without legerdemain [sleight of hand]."

Adrift . . . (loose, unmoored)

A ship is said to be "adrift" when she is without a mooring and is being carried along at the mercy of the sea, much like society's drifters who are carried from place to place by the tidal forces of life, without direction or purpose. The philosopher John Locke wrote in his *Essay on Human Understanding,* "and so we should let our thoughts (if I may so call it) run adrift without Direction or Design." The word derives from the Middle English *drifte* (to float).

Ahoy! . . . (nautical salutation)

Considered by some authorities to have been an ancient Viking battle cry, the interjection is now commonly used to hail a ship. Alexander Graham Bell suggested *Ahoy!* as the appropriate salutation for answering the telephone. Although his recommendation was not accepted by society at large, the offshoots *Hi!*, *Ho!*, and *Yo!* have crept into colloquial English. From historical fiction to the music hall, Victorian writers used nautical *Yo-ho*'s to add atmosphere and salt to their characters and skits. Operettas by Sir William S. Gilbert and Sir Arthur Sullivan are liberally sprinkled with such derivatives. The well-known refrain from *The Mikado* is a case in point:

But the happiest hour a sailor sees
Is when he's down
At an inland town,
With his Nancy on his knee, yo ho!
And his arm around her waist!

Albatross . . . (encumbrance or handicap)

The albatross is a long-winged seabird held in reverence by mariners as an omen of good luck. In Samuel Taylor Coleridge's narrative poem *The Rime of the Ancient Mariner,* the title character committed an unspeakable act by killing an albatross that had guided the ship through treacherous fog and floating ice. When all manner of misfortune descended upon the vessel, the desperate crew draped the dead albatross around the Ancient Mariner's neck as a sign that he, alone, bore the burden of guilt for its death. Eventually, after many hardships and much suffering, the penitent Ancient Mariner was forgiven and died "a sadder and wiser man." In colloquial English, however, anything that is cumbersome, causes deep concern, or makes difficult the accomplishment of a goal is said to "hang around one's neck like an albatross."

All at Sea . . . (confused)

The expression is an allusion to the uncertain plight of a ship drifting about aimlessly, unable to find her bearings on the vast, open sea. An individual is said to be "all at sea" when in a state of intellectual or emotional confusion. Selous, in his 1893 *Travels in Southeast Africa,* noted, "I was rather surprised to find that he seemed all at sea, and had no one ready to go with me."

All in a Day's Work . . . (routine)

William Falconer, in his reference book *Universal Dictionary of the Marine* (1789),
defined a day's work as "the reckoning or account of the ship's course during
the twenty-four-hour period between noon and noon." The expression has
crept into modern English usage as the normal course of events that occurs
in any regular work period.

Aloof . . . (removed, at a distance)

In nautical terminology, the word *aloof,* which derives from the Old Dutch *loef* (windward), means to sail a ship to windward, thereby avoiding a lee shore. A
vessel that maintains a position to windward of other ships in her company
is said to "stand aloof." Accordingly, *aloof* describes an individual who is
reserved and keeps at a distance. In 1817, Samuel Taylor Coleridge wrote
in his opus *Biographia Literaria,* "the alienation, and if I may hazard such
an expression, the utter aloofness of the poet's own feeling."

Anchor . . . (base of strength, symbol of hope)

Technically, an anchor is a device for holding a ship in a particular place. Metaphorically,
any reliable support or
base is considered to be
an anchor. As Alfred,
Lord Tennyson noted,
"Cast all your cares on
God; that anchor holds."
In the Christian religion
the anchor is a symbol
of hope: "Which hope
we have as an anchor
of the soul, both sure
and steadfast . . ."
(Heb. 6:19).

Antenna . . . (projecting rod or feeler)

In ancient Greek, the word *anateinein* (to stretch forth) described the projecting horns
of insects. Fifteenth-century Latin writers, looking for a word to describe
the long, high-peaked yard on a lateen sail, and noting its resemblance
to an insect's projecting horn, borrowed from the Greek and coined the
word *antenna*. Modern English retains the Latin spelling, and the Greek
definition, and adds a meaning that would have had little relevance to the
ancient Greeks or Latins—today, an antenna is also a projecting metal rod
for receiving radio signals.

Any Port in a Storm . . . (anything will suffice when in need)

The nautical origin of this popular saying is obscure. It first appeared in *The Pirate*, a
sea adventure written in 1836 by British naval officer and novelist Captain
Frederick Marryat. The author may have observed the popularity of the
expression among sailors, or he may have invented the phrase to add flavor
to his story. In any case, a nautical allusion is inherent in the expression
any port in a storm.

Armed to the Teeth . . . (well prepared)

A product of Victorian embellishment, the expression is sheer romanticism. In a speech
to Parliament in 1849, English statesman Richard Cobden asked his
colleagues, "Is there any reason we should be armed to the teeth?" It is
not known whether Cobden invented the expression or merely repeated
one that had already been introduced to the realm of colloquial English.
Hollywood—primarily with the production of such films as United Artists'
The Sea Hawk—confirmed and preserved for subsequent generations the
Victorian image of the daring buccaneer liberally armed with an assortment
of weapons. This colorful colloquialism has crept into common usage as a
reference to anyone who is well prepared and appropriately equipped.

Avast! . . . (stop!)

This interjection is the order to stop or hold in any nautical operation. Some authors cite the Italian *basta* (enough) as the word's origin. The *Oxford English Dictionary,* however, cites the derivation as Old Dutch *houd vast* (to hold fast). In either case, the once strictly nautical command is now used colloquially as an instruction to cease and desist any action or behavior. British naval surgeon and writer Tobias Smollett was among the first to use the word colloquially. In his 1748 novel *The Adventures of Roderick Random,* one of his characters says, "Avast there, friend, none of your tricks upon travelers!"

Awash . . . (helpless)

Literally, *awash* means covered with water. The nautical usage refers to the condition of a ship when she is almost submerged and at the mercy of the sea. Figuratively, the term is used frequently to describe an individual so inundated by something that regaining control or understanding of the situation is difficult. It is like being awash in a sea of hyperbole during a long-winded sermon.

Bad Name . . . (unsavory reputation)

In *The Sailor's Wordbook* (1867), Admiral William Smyth advised mariners to guard their own reputations and that of their ship, "for once [a bad name is] acquired for inefficiency or privateering habits, it requires time and reformation to get rid of it again." A similar situation exists on shore. Society tends to equate an individual's worth with a good or bad reputation. The latter is a black mark difficult to overcome; the former is a sign of credibility that carries with it the suggestion of immortality. In Shakespeare's history *King Henry IV, Part I,* when Harry Hotspur realizes he has been outmatched in battle, his greatest concern is the damage to his reputation rather than his loss of life: "I better brook the loss of brittle life / Than those proud titles thou hast won of me," he moans just before dying.

Ballast . . . (stability)

This term comes from the Old Teutonic *ballast* (belly load). Nautical usage of the word refers to weight (other than cargo) carried deep within the hold, or "belly," of a ship to give her stability and trim in the water. After a ship discharges her cargo, additional ballast is taken on to compensate for the loss of weight. Literally and figuratively, the word *ballast* is used to convey a sense of stability to something or someone. In *A Christian in Complete Armour*, seventeenth-century English divine William Gurnall admonished, "If he be not well ballast with humility, a little gust will topple him into sin."

Barge In . . . (to interrupt in an abrupt manner)

The word *barge* has two totally different meanings. The first stems from the Latin *barca* (boat), a richly decorated state vessel propelled by oarsmen for ceremonial occasions and other pageants. William Shakespeare described Cleopatra's barge as being "like a burnished throne /. . . Purple the sails, and so perfumed, that / The winds were love-sick with them. . . ." The second meaning is more commonplace: a large, unwieldy but functional, flat-bottomed boat used primarily for hauling freight in rivers and inland waterways and, in recent years, for deep-sea towing. Barges were once pulled through canals by mules on the bank. A traditional American folk song dating from the 1850s begins with the line, "I've got a mule, her name is Sal, fifteen miles on the Erie Canal." Propelled by "mulepower," barges could cover long distances but were clumsy, difficult to control, and frequently involved in collisions with other craft. It was Sal's cumbersome canal barge, not Cleopatra's burnished throne, that gave rise to the colloquial expression "to barge in," that is, to make an abrupt or un-welcome interruption. To para-phrase Alexander Pope, "Fools barge in where angels fear to tread."

Batten Down the Hatches . . . (prepare for a storm)

The Oxford Companion to Ships and the Sea defines a *hatch* as "an opening in the deck of a ship used for ingress and egress of personnel and cargo." A nautical batten is a thin wooden strip used primarily to secure tarpaulins over the hatches. The process of battening down the hatches—done as a precaution against rough and stormy weather—involves securing canvas over the wooden boards that cover the hatches. Even among landlubbers, who rarely go down to the sea in ships, this salty metaphor is frequently borrowed to describe preparing for a stormy situation ashore, whether figuratively or literally.

Beam Ends . . . (near ruin)

A ship is in imminent danger of sinking when she heels over so far she may not be able to regain her normal, upright position. In this condition, her deck beams are almost perpendicular to the water's surface and she is said to be on her "beam ends." A sailor who is on his beam ends is flat broke and at a loss for any prospect to right himself. In 1844, Charles Dickens used the expression in his novel *Martin Chuzzlewit:* "Tom was thrown upon his beam ends again for some other solution."

Belay . . . (stop)

In nautical parlance, to *belay* is to take turns with a rope around a cleat, fasten it, and make it secure. In a traditional sea chanty, sailors sing "I thought I heard the old man say, give one more haul and then belay." On land or at sea, *belay* is a general order to stop or hold. As William Smyth noted in *The Sailor's Wordbook,* "Belay there, stop, that is enough. Belay that yarn, we have had enough of it."

Bent on a Splice . . . (amorous union)

Splicing is the procedure of uniting two ropes or two parts of the same rope by intertwining the individual strands. A sailor or landlubber who is *bent on a splice* is one who is about to be united to his lady-love within or without the bonds of matrimony.

Betwixt Wind and Water . . . (in a vulnerable spot)

This expression refers to an area close above and below the waterline on a wooden ship, a space exposed alternately to air and water as the ship rolls. In the days of wooden warships, a vessel taking a hit from an enemy cannon in this vulnerable spot was in danger of sinking as water gushed in through the shot hole when she rolled. As William Robinson wrote in his book *Jack Nastyface, Memoirs of an English Seaman,* "Some of our men were sent on board of the Spanish ship before alluded to, in order to assist at the pumps, for she was much shattered in the hull, between wind and water." Born at sea, this expression has washed ashore where it is generally used to describe being hit, figuratively or actually, in a way that inflicts significant damage.

Bigwig . . . (person of importance)

Seen in print as early as 1792, *bigwig* has its origins in the large wigs once worn by men of rank and distinction. The term formerly carried the connotation that, although the wigs themselves were of formidable size, the heads they covered had little or nothing in them. Smyth defined "bigwig" in *The Sailor's Wordbook* as an expression "applied to high-ranking officers." Modern social usage is usually applied to VIPs, CEOs, and other individuals who have inordinate wealth or position.

Bilge . . . Bilge Water . . . (nonsense)

The bilge is the bottom of a ship's hull. Seawater, rainwater, and all kinds of waste matter ultimately collect in the bilge. The composite, called *bilge water,* breeds an offensive smell. Richard Henry Dana's *Two Years Before the Mast,* the classic adventure tale based on his own experiences at sea in the 1830s, describes the "inexpressibly sickening smell caused by the shaking up of the bilge water in the hold." Sailors were known to become mortally ill from the foul-smelling fumes and gases emanating from the bilge. Ashore, the mention of "bilge" or "bilge water" conveys a sense of garbage, nonsense, and poppycock.

Bitter End . . . (to carry a long, difficult struggle to its inevitable conclusion)

The usually accepted explanation of the origin of this popular metaphor has a distinctly nautical origin. The anchor rope (which today is called "line") on old sailing vessels was attached to a stout oak post called a *bitt,* which was firmly fastened to the deck. Securing turns were taken around the bitt as anchor and anchor rope were paid out to the sea. The end of the rope nearest the bitt was called the *bitter end.* When at the end of your rope, on land or at sea, you've reached the bitter end.

Black Book . . . (record of those in disgrace)

The ancient Laws of Oleron, named for the Island of Oleron, a famous seafaring community in the Duchy of Aquitaine, were introduced into England during the waning years of the twelfth century. Essentially a collection of maritime laws and traditional customs of the sea, they were codified in 1336 under the title *Rules for the Office of Lord High Admiral; Ordinances for the Admiralty in Time of War; the Laws of Oleron for the Office of Constable Marshall; and Other Rules and Precedents.* It addresses, among other things, the responsibility of a ship's captain to enforce discipline and mete out punishment. Because the document was bound in black leather, or perhaps because its official title was such a tongue twister, it became known as the *Black Book of the Admiralty.* The original *Black Book* no longer exists, having "disappeared" from the High Court of the Admiralty around the turn of the nineteenth century. But, according to remaining fragments of manuscript copies, as well as to writings of eighteenth-century scholars, punishments laid down in the *Black Book* were unspeakably cruel. For serious offenses such as repeatedly sleeping on watch, shipboard pilfering, theft, and murder, punishments included drowning, starvation, and marooning. Thus, "black book" became associated with misdeeds, crime, and punishment. Colloquially, to be listed in someone's black book means simple disgrace as a result of another's displeasure—today, though, without the dire consequences cited in the original *Black Book.* The concept of a black book intrigued Elizabethans. The poet Edmund Spenser, who was probably listed in a number of black books, referred to what must have been an accepted practice: "all her faults in thy black book enroll."

Blazer . . . (informal sports jacket)

The word *blazer* has its origins in the Old English *blaese* (a bright torch or firebrand), and there are several popular theories about its nautical origin. One possibility is a link to the blazing scarlet jackets that were first worn in 1889 by the crew of the *Lady Margaret* at the Boat Club of St. John's College, Cambridge. Another theory centers on a nineteenth-century British naval custom that permitted captains to buy distinctive jerseys for their crews. The colorful results of this custom have become legendary. Some authors have cited the nineteenth-century H.M.S. *Blazer* as being the nautical source of the word because her crew turned out in striking blue-and-white–striped jerseys. The blazer-jacket prototype was a combination of bright team or club colors and was more reminiscent of a storefront awning than of the conservative, navy blue wool and flannel garment worn today. Blazers appear to have come into vogue in England during the Victorian era. Articles in various newspapers and journals during the early 1880s refer to "men in spotless flannels and club blazers," and to the latest novelty and fashion on the river being "blazer and spats."

Blood Is Thicker Than Water . . . (loyalty because of a blood relationship)

When the British were defeated during a land attack on the Peiho forts during the Second China War in June 1859, Commander Josiah Tatnall of the U.S. Navy came to their aid. By towing boatloads of British survivors from shore, Tatnall violated the neutrality of the United States. The act might have led to serious diplomatic and political repercussions had the commander not succeeded in justifying his behavior with the argument that *blood is thicker than water.* The Georgia-born Tatnall may have held the same sentiment at the outbreak of the Civil War in 1861, when he accepted an appointment as senior flag officer in the Confederate Navy. Later, after the evacuation of Norfolk in May 1862, in order to prevent the federal forces from capturing his flagship, the ironclad C.S.S. *Virginia*, Tatnall burned and sank her. That was her second trip to the bottom of the sea, and for a similar reason. The C.S.S. *Virginia* was none other than the U.S.S. *Merrimac,* sunk by the U.S. Navy in April 1861 to prevent her capture by the Confederates. However, her maiden voyage to the briny deep was all for naught. The Confederates raised, rebuilt, renamed, and deployed her against her former owners. Northern journalists of the period, as well as subsequent generations of writers, clung doggedly to the use of C.S.S. *Virginia*'s original name. As a result, the engagement between the two ironclads remains known as the battle between the *Monitor* and the *Merrimac.* Family ties in the Civil War notwithstanding, for the *Merrimac,* water proved thicker than blood.

Blue Monday . . . (dreary day)

In his book *A Sea Grammar,* Captain John Smith described the following punishment for sailors caught telling a lie: "The liar is to hold his place but for a week, and he that is first taken with a lie, every Monday is so proclaimed at the main mast by a general cry, 'A Liar, A Liar, A Liar . . . he is under the Swabber, and only to keep clean the beak-head and chains.'" (See also "Head" and "Swab.") In 1685, Captain Nathaniel Boteler, known for his interesting accounts of naval rules and customs, noted, "the idleness of ships' boys is paid out by the boatswain with a rod . . . and commonly this execution is done on Monday mornings." This practice of meting out punishment on Mondays is said to be the origin of the popular expression *blue Monday,* a reference to the dreary day that marks the beginning of another workweek.

Bonanza . . . (prosperity)

The word *bonanza* is derived from the Spanish *bonansa* (to sail with fair wind and weather). It is thought that American mariners borrowed the word from their Spanish counterparts and used it to describe prosperity in general. In turn, gold rush miners heard the word aboard ships bound for California and used it to describe any rich find of precious metal. The word crept into colloquial American English and has come to mean anything that is considered abundant, even as R. Taylor mused in an 1878 article in the *North American Review*, "if silence be golden, he was a bonanza!"

Booby Hatch . . . (mental institution)

Some etymologists give a nautical derivation for this insensitive colloquial expression. They cite the practice of punishing sailors by confining them in the booby hatch—a small, hooded compartment located near the bow of the ship. The term is said to have arisen as a result of the screams of the unfortunate sailors imprisoned in the cramped, stifling confines of the booby hatch. It is more likely, however, that the slang connotation arose from the word *booby* itself. *Booby* has its origins in the Spanish word *bobo* (slow-witted and foolish). In 1634, the celebrated author and traveler Sir Thomas Herbert described a tropical bird that perched on the yards of ships and allowed itself to be caught easily: "one of the sailors espying a bird fitly called a Booby, he mounted to the topmast and took her. The quality of which bird is to sit still, not valuing danger." It is easy to imagine the fun and diversion that bored, lonely sailors had in catching these dimwitted birds by hand, and confining them in a small, hooded coop called a *hutch,* a word easily corrupted to *hatch.*

Bootleg . . . *(to sell or traffic in goods illicitly)*

Strictly American, this term is thought to have originated from the practice of sailors who once smuggled goods ashore in the upper part of their seaboots. (See also "Real McCoy.") The word washed ashore and today may even be found on the football field, where *bootleg* describes a play in which the quarterback fakes a handoff to a teammate, conceals the ball on his hip, and then runs with it.

Born with a Silver Spoon in His Mouth . . . *(privileged)*

During the nineteenth century, some young men whose families had pedigree or connections were allowed to enter the Royal Navy without taking the requisite examination. Their subsequent promotion was taken for granted. The rank-and-file sailor referred to these wonder boys as having entered the navy through the stern cabin windows. Because *silver plate* was used for meal service in both officers' cabins and in the homes of the well-to-do—a reference to the elite—the phrase soon evolved to the colloquialism *born with a silver spoon in his mouth*. Less privileged and, therefore, lower-class sailors were referred to as having entered the navy through a hawsehole (an opening on the bow of a ship through which anchor lines pass). The ordinary sailor, in direct contrast to his more fortunate shipmate, was referred to as having been *born with a wooden ladle in his mouth*. His promotions were, of course, based on merit, not family status. Although the descriptive tags have all but dropped from modern usage, the expression *silver spoon* has washed ashore and, in a general sense, refers to anyone who is wealthy, well connected, or privileged.

Brace of Shakes . . . *(very short period of time)*

The word *brace*, meaning two or a pair, derives from the Old French word *brasse,* meaning the width of two arms. A *brace of shakes* is an old nautical term that refers to the fraction of a minute's time in which a shaking movement could be detected in the sails as a ship was brought into the wind. In

Robert Louis Stevenson's classic tale *Treasure Island,* Captain Smollett remarked, "we'll be going in two shakes," or in as short a period of time as it takes for the sail to shake twice.

Braced Up . . . (under control)

In a square-rigged ship, *braces* are two opposing sets of lines that control the swing of the yards, and from which sails are hung. A sail is "braced up" when it is drawn taut in order to sail as close to the wind as possible. Ashore, an individual is said to be "braced up" when tensed in preparation for a figurative or literal blow. In Leigh Hunt's 1847 collection of essays, memoirs, and uncollected prose entitled *Men, Women, and Books,* he used the expression in its colloquial form: "would to heaven his nerves had been as braced up as his face."

Brought Up Short . . . (unexpected standstill)

In the days of sail, a vessel underway could be brought to an emergency standstill, or brought up short, by dropping the anchors. The sight and sounds made by a 2,000-ton, heavily armed frigate grinding to a sudden halt against the drag of anchors must have been a nerve-wracking experience—particularly to those on board. Metaphorically, a person is *brought up short* when forced to a standstill by the drag of a figurative anchor in the form of a sudden, inconvenient reversal of fortune.

Buccaneer . . . (reckless adventurer)

From the French word *boucane* (a small smokehouse or grill), *boucaniers* were men who hunted and smoked game. Over the years, the term was broadened to include not only the men who hunted and prepared the meat, but those who ate it as well. Caribbean pirates became known as "buccaneers" because they bought great quantities of smoked meat to eat at sea during their long, marauding expeditions. When buccaneer-surgeon John Esquemeling's *Bucaniers of America* was translated into English in 1684, it became an international best-seller. His classic account of buccaneering adventures has continued to inspire writers into the present century, including English Poet Laureate John Masefield. (Masefield's poem *The Tarry Buccaneer* is a fine example.) The word *buccaneer* even penetrated the world of English politics and government—in 1877 William Gladstone said in a speech, "some of our less temperate adventurers (I must not call them buccaneers)." The American poet Emily Dickinson used the word whimsically when she referred to bees as "Buccaneers of Buzz" in her poem "Bees are Black, with Gilt Surcingles—."

Buoy . . . (encourage)

A buoy is a floating navigational aid that marks shoals, channels, or other areas of concern for navigators. The word probably has its origins in the Spanish *boyar* (to float) and/or the Old French *boyee* (to fetter). Although buoys are affixed to the seabed by a chain or cable, they appear to bob and float freely on the surface of the water. Figuratively, then, *to buoy up* means to uplift or to sustain by encouragement, cheerfulness, or other positive expressions. In *Dante and His Circe*, Dante Gabriel Rossetti observed, "The spirits of thy life depart daily to heaven with her so they are buoyed with their desire."

By and Large . . . (for the most part)

To sail a vessel *by and large* means to sail her as close as she can go to the wind without being hard on it. On shore, when a person doesn't wish to "sail" directly into a subject, *by and large* serves as a circumspect expression meaning "for the most part." In the British Broadcasting Company's highly acclaimed screen adaptation of Jane Austen's novel *Pride and Prejudice*, Miss Mary Bennett remarked on the betrothal of Mr. Wickham to the freckled, odious Miss King: "by and large, it was to be expected."

By Guess and by God . . . (inspired guesswork)

Sailors have long been aware of a divine power that shapes their fragile destinies on vast and lonely seas. William Shakespeare interpreted their prayers to the Almighty in *Pericles*:

> *Thou God of this great vast, rebuke these surges,*
> *Which wash both heaven and hell; and Thou hast*
> *Upon the winds command, bind them in brass,*
> *Having call'd them from the deep. O, still*
> *Thy deafening, dreadful thunders; gently quench*
> *Thy nimble, sulphurous flashes.*

Although not expressed with Shakespeare's genius, simple faith in divine power was reflected in many aspects of day-to-day routine aboard ship. Sails and fittings were nicknamed "Trust-in-Gods," "Hope-in-Heavens," "Apostles," and "Paternosters." One form of navigation, known as *by guess and by God,* relied on a combination of educated guesswork, the intuition of the ship's master, and the good graces of the Almighty. When this expression is used on land, it means arriving at a solution or conclusion by intuition and good luck.

By the Boards . . . (missed opportunity)

When a ship's mast falls over the side and is carried away, it is said to have gone by the boards or, literally, by the wooden deck and hull planking. Figuratively, the expression means something that has passed by, particularly a missed opportunity.

Caboose . . .
(end car on a freight train with eating and sleeping facilities for its crew)

Although its etymology is obscure, the term *caboose* was born at sea and is thought to be from Dutch or Low German. A ship's caboose was a cook room or diminutive galley, introduced into the Royal Navy during the mid-eighteenth century. Unlike traditional galleys, which were located between decks, the caboose was located on an open deck of the ship. In his *Universal Dictionary of the Marine* (1789), William Falconer described the caboose as "a sort of box." While the nautical application of caboose is now obsolete, today *caboose* describes the galley-sleeping car used by freight train crews.

Career . . . (to tip and sway to one side while in motion)

This term comes from the Latin *carina* (keel). To career a ship is to heave her down deliberately on one side, so that the other side comes out of water for repairs and maintenance such as caulking, scraping, or plugging shot holes. Since the introduction of dry docks and hydraulic lifts, which hoist the entire hull of a ship out of water, the procedure of careening a ship has become almost obsolete. The word, though, lives on in colloquial English. In his historical romance *Ben Hur*, General Lew Wallace noted, "the charm of the camel is not in the movement, the noiseless slipping or the broad career."

Chewing the Fat . . . (idle gabbing)

In the old days of wooden ships and iron men, crews talked and grumbled while "chewing the fat," their daily ration of brine-toughened salt pork. *Chewing the fat* is a nautical expression that lost its negative overtones when it washed ashore. It has come to mean an idle, friendly conversation.

Chit . . . (a signed voucher for goods or services)

From the Hindi *chitti,* meaning a letter or note, *chit* came into popular usage during the heyday of British ships sailing to and from India. In a 1774 issue of the *Indian Observer*, Hugh Boyd bemoaned the practice: "The petty but constant and universal manufacture of chits which prevails here."

Chock-Full . . . Chuck-Full . . . (filled to the extreme limit)

The expression "chokefull charged with gold"—that is, so full as to cause choking in a figurative sense—was transmitted to the English-speaking world in the late fifteenth century by Thomas Malory in his book *Morte D'Arthur*. Probably a corruption of the word *choke,* a *chock* is a wooden wedge that stabilizes cargo in a ship's hold and keeps it from shifting while the vessel is underway. The expression *chock-full* alludes to a ship's hold that is filled to capacity. Naval surgeon Tobias Smollett offered this advice in his *Adventures of Peregrine Pickle:* "stow thyself choque-full of the best liquor in the land."

Chow . . . (food)

This is another word that dates back to the days of clipper ships and the tea trade to India and China. It is thought to be a shortened version of the Chinese-pidgin word *chow-chow,* which means food.

Clean Bill of Health . . . (healthy, in good shape)

This popular expression derives from the certificate once issued by a port authority confirming that no member of a ship's crew suffered from a reportable contagious disease, and that no contagion was known to be present in the ship's port of departure. Where there was infectious disease aboard a ship or in port, the authorities issued a foul bill of health. As the term "a clean bill of health" has come ashore, it generally means "in good shape," as when a company is given a clean bill of health on its financial dealings.

Clean Slate . . . (fresh start)

In the days of sail, the courses and distances made good during each watch were temporarily recorded on a slate. After transferring the information into the ship's log, the slate was wiped clean prior to the next watch. Colloquial use of this expression ashore means to forget past events and start from "scratch."

Close Quarters . . . (immediate contact)

In nautical lingo, *close quarters* were wooden barriers erected across the decks of merchant ships. Fitted with *apertures* (loopholes) through which small arms could be fired, *close quarters* afforded a place of protection or retreat when the ship was boarded by pirates. In its colloquial application ashore, *close quarters* refers to something done hand-to-hand or at close range. Thus, a jogger might come into close quarters with a snarling Doberman. The expression also denotes an attitude of intense scrutiny, as in coming into close quarters with one's own conscience or values.

Cock-Up . . . (foul-up)

When the yards (horizontal spars) on a square-rigged ship are lifted up so that they lie at an angle to the masts (vertical spars), they are said to be "cocked-up" (turned up), or "cockbilled." Yards are also said to be "scandalized" (from the Latin *scandalum,* a cause of offense) when they are cocked-up. Yards were often cockbilled so that a ship could lie close alongside a building by a pier. They were also cockbilled as a sign of mourning for the death of a crewmember, but other than when done deliberately for one of those purposes, having the yards accidentally cockbilled was considered sloppy practice. The colloquialism *cock-up,* which means a hopeless, messy foul-up, is now heard frequently on television programs originating from the British Broadcasting Company. Although the expression itself is not a scandal or cause for offense, the situation it describes might well be.

Cold Enough to Freeze the Balls off a Brass Monkey . . . (extremely cold temperature)

As a child I was admonished by my mother not to repeat anything that my sailor uncles said. One frigid morning, I was pleasantly startled to hear my eldest uncle, lately returned home from the Brooklyn Navy Yard, exclaim, "Cryin' out loud! It's cold enough to freeze the balls off a brass monkey!" Thinking that it sounded more naughty than nautical, my mother looked appropriately horrified. Although I was dying of curiosity, my mother's reaction warned me not to ask the obvious. Some years later, when I was a young student nurse, one of my patients was the unforgettable Captain Hezekiah Litchfield, a ninety-two-year-old Maine native who had spent almost his entire life at sea. Complaining vehemently about the lack of heat in the large, open ward, the curmudgeonly old sailor bellowed, "[expletive deleted] It's cold enough to freeze the [expletive deleted] balls off a [expletive deleted] brass monkey." After I gave him a new, woolen blanket, I mustered enough courage to ask the question that had been plaguing me for years. With the steamier sides of the old salt's vocabulary either deleted or sanitized, Captain Hezekiah Litchfield's story follows.

When tall ships and great sailing navies dominated the seas, a first-rate ship of the line could carry as many as one hundred heavy cannon along with many smaller ones—but the cannon balls occupied valuable space aboard the cramped vessel. To solve the problem, balls were stacked pyramid fashion on brass trays called "monkeys." During long periods of very cold weather, the balls would shrink (thermal contraction), shift, and fall off the brass monkey. (The same basic story was related by Bill Bevis and Richard McCloskey in their 1983 book *Salty Dog Talk*.) Because many sailors could neither read nor write in the bygone era of sail, oral tradition has played an important role in perpetuating the rich and colorful language of the sea. Whether this largely unsubstantiated tale about balls and monkeys is true or apocryphal, the saying has entered the realm of modern colloquial usage. The Italian adage *se non è vero, è bene trovato*—it's not the truth, but it's well invented—may well apply here.

Copper-Bottomed Investment . . . (secure, sound)

In 1761, the Royal Navy introduced the practice of sheathing the hulls of wooden ships with copper to prevent the teredo worm (the equivalent of an oceangoing termite) from eating into planks beneath the waterline. It was an expensive process but one that proved to be a significant development in shipbuilding. The term *copper-bottomed* has come to mean an investment that is strong and secure, one that can be trusted.

Cranky . . . (irritable)

In *A Sea Grammar*, Captain John Smith described a "crank-sided" ship as one that heels over too easily. Probably from the Dutch-Frisian word *krengd*, (laid over on its side), the term *cranky* describes a ship with low stability. Poet-diplomat John Russell Lowell seemed to be preoccupied with nautical metaphors when he wrote, "There is no better ballast for keeping the mind steady on keel, and saving it from all risk of crankiness, than business."

Cut a Dido . . . (dashing)

H.M.S. *Dido* was a swift, armed light cruiser commissioned in the Royal Navy during the late eighteenth century. Apparently her captain and crew loved to show off their ship, as well as their own prowess, by cruising smartly around other ships in the fleet. Having literally "sailed rings around" them, H.M.S. *Dido* then dropped anchor. When one *cuts a Dido* ashore, it means to create a sharp, dashing appearance.

Cut and Run . . . (hasty departure)

This expression dates back to the days of square-rigged sailing ships when, on occasion, the urgent need to get under way could be anticipated. In preparation, sails were secured to the yards with light ropeyarn that could be cut to let the canvas fall quickly, thus enabling the ship to sail at once. Figuratively, one *cuts and runs* when making an unceremonious and hasty departure.

Cut of His Jib . . . (outward appearance)

A jib is a triangular sail set on the stays of the foremast. In the days of ships under sail, the cut of a jib indicated the type of a ship and in some instances, her nationality. The expression, "I don't like the cut of his jib," translates to judging a person by his appearance or outward demeanor and taking an instant dislike to him.

Cutting the Painter . . . (sneaky departure)

A nautical painter is a length of small rope by which a boat is secured to a pier, buoy, dock, or to a ship itself. The expression *cutting the painter* means making a getaway or a clandestine departure, for when a boat's painter is cut, the boat can drift silently away. In a sailor's lingo, since it is the painter that secures a small boat, *cutting the painter* also means to sever one's lifeline or to die, the landlubber's equivalent of *buying the farm*.

Davy Jones's Locker . . . (bottom of the sea)

The origin of this expression remains obscure. Some etymologists believe that it is a corruption of "Duffy Jonah," an expression used by West Indian sailors in reference to the devil. Others believe that "Davy" derives from St. David, the patron saint of Wales who was often invoked by Welsh sailors. Another school of thought believes that "Jones" is a corruption of "Jonah," the name of the Old Testament prophet who was swallowed by a whale and spewed back up on land after three days. The phrase came into popular usage more than two centuries ago and, in nautical parlance, refers to a spirit of the deep, often but not always malevolent. British naval surgeon and novelist Tobias Smollett first described Davy Jones in his 1751 novel *The Adventures of Peregrine Pickle:*

> *I'll be damned if it was not Davy Jones himself. I know him by his saucer eyes, his three rows of teeth, and tail, and the blue smoke that came out his nostrils. This same Davy Jones, according to the mythology of sailors, is the fiend that presides over all other evil spirits of the deep, and is often seen in various shapes, perching among the rigging on the eve of hurricanes, shipwrecks, and other disasters to which seafaring life is exposed, warning the devoted wretch of death and woe.*

◆ ◆ ◆

According to nautical tradition, *Davy Jones's Locker* (the locker being an allusion to a sailor's trunk or sea chest) is the final resting place of sunken ships, articles swept overboard, and, of course, those buried or lost at sea.

Dead Marine . . . (empty bottle)

Dead marine or *dead soldier* refers to an empty wine bottle. The expression has been attributed to William IV, England's highly eccentric Sailor King (1830–1837). It was said that William, while still the Duke of Clarence, ordered a ship's steward to remove the "dead marines" (empty wine bottles) remarking that, like the marines, the wine had done its duty nobly and would be ready to do it again. Edward Trelawny used the expression in print as early as 1831 when he advised a younger son "to see their cases properly filled, no marines among them." But sailors had scant respect for the marines, who were but landsmen; that is, soldiers specially trained for warfare at sea. Cynical old salts soon gave the expression the implied meaning that an empty bottle was as useless as a dead marine, live ones being useless enough. Paradoxically, both the United States Marines and the British Royal Marines are now recognized as the elite among fighting forces.

Deep Six . . . (to get rid of)

To *deep six* is a sailor's expression for throwing something overboard. The "six" refers to the 6-foot nautical fathom, the standard unit of measurement for sea depth. In contemporary usage ashore, the phrase now means to kill something, or to send it to the bottom, as in "deep sixing" a request for funds.

Deliver a Broadside . . . (verbal assault)

In *The Sailor's Wordbook* (1867), Admiral William Smyth defined a broadside as "the whole array, or the simultaneous discharge of all the artillery on one side of a ship of war." Considering the size and firepower of vessels of the period, two warships firing broadsides at each other must have been an appalling spectacle of death and destruction. An early-nineteenth-century street ballad entitled "Nelson's Death and Victory" describes the Battle of Trafalgar:

> *Broadside and broadside our cannon balls did fly,*
> *And smallshot like hailstones on the deck did lie,*
> *The masts and the rigging were all shot away.*
> *Besides, some thousand on that day*
> *Were killed and wounded in the fray*
> *On both sides, brave boys.*

When a broadside is delivered ashore, it means to fire a heavy volley of abuse or denunciation. By land or by sea, heed the words of William Shakespeare: "Fear we broadsides? No, let the fiends give fire!" *(King Henry IV, Part II).*

Derelict . . . (noun: something run-down and abandoned; adjective: negligent)

From the Latin *derelinquere* (to forsake), a ship is said to be a derelict when she is abandoned at sea, whether by choice or necessity. A person who is down on his luck and has been abandoned by society is also considered a derelict. On shore or at sea, a person is derelict in duty if neglectful of the obligations that it imposes. In his 1840 book *The Tower of London,* William Harrison Ainsworth remarked that "they would be answerable with their lives for any further dereliction of duty."

Devil to Pay . . . (facing serious consequences)

Between the Devil and the Deep Blue Sea . . . (caught in a tight spot)

Aboard wooden sailing ships, the *devil* was the name given to the seam formed at the juncture where the covering board that capped the ship's sides met the deck planking. The seam was particularly difficult to caulk because of its length, because there was so little space in which to perform the awkward task, and because there was so little standing between the *devil* and the sea. (The caulking was sealed by pouring hot pitch into the seam—a process known as "paying.") Although aboard ship the complete expression was "There'll be the devil to pay and only a half bucket of pitch," being *between the devil and the deep blue sea* literally means being on the narrow plank (the covering board) with little room to maneuver. This colorful expression now means being caught in a tight spot with few options.

In coming ashore, both expressions lost their original meanings as landlubbers assumed that the devil was Satan. Thus, "the devil to pay" has evolved to mean "to catch hell," and "between the devil and the deep blue sea" means "no good option." Helping to keep these colorful devils alive today is Ella Fitzgerald's rendition of a hopeless love-hate relationship in Harold Arlen and Ted Koehler's classic song "Between the Devil and the Deep Blue Sea."

Dirty Dog . . . (a worthless, dishonorable person)

The original nautical expression, "a dirty dog and no sailor" was defined by Admiral William Smyth in 1867 in *The Sailor's Wordbook* as "a mean, spiritless and utterly useless rascal."

Ditty Bag or Box . . . *(utility bag for small tools or personal effects)*

In *The Sailor's Wordbook,* Admiral William Smyth explained that the ditty bag "derives its name from *dittis* or Manchester stuff of which it was once made." In the nineteenth century, Manchester, England, was renowned for its cloth industry. Presumably, *dittis* was a kind of duck or canvas material made in the area, but nothing further is known about it. In his 1860 *Seaman's Catechism,* Stuart described the contents of a typical ditty bag: "to contain two dozen of clothes, stops, needles, thread, scissors, tape, buttons and thimbles." Despite the obscurity of the phrase's origin, ditty bags have accompanied sailors to sea for centuries.

Doldrums . . . *(boredom)*

The belt of calm lying close to the equator is known as the doldrums. Sailing ships that cross the doldrums roll in the ocean swells with their sails slatting uselessly, catching whatever breeze they can in fits and starts—an annoying sound and a very frustrating time for sailors. Figuratively, an individual is said to be "in the doldrums" when intellectually stagnant and bored. The colloquial expression is thought to be an analogy to the physical and mental state of the sailors whose ships were becalmed in the stifling latitudes of the doldrums. As Sir Thomas Sutherland wrote in 1895 in the *Westminster Gazette,* "The ship of State has escaped the tornado but seems becalmed in a kind of political and financial doldrums."

Don't Give Up the Ship . . . (don't lose heart, keep going)

In June of 1813, Captain James Lawrence, in command of the U.S.S. *Chesapeake*, engaged the British frigate H.M.S. *Shannon* just outside Boston Harbor. After a short, bloody battle, the *Chesapeake* was seriously damaged and her captain lay mortally wounded. Reportedly, Lawrence died with his last command still on his lips: "tell the men to fire faster . . . fight 'til she sinks, boys . . . don't give up the ship." The Americans lost the battle and were compelled to surrender the *Chesapeake,* but Lawrence's dying words lived on. Commodore Oliver Hazard Perry, who is frequently and incorrectly credited with being the source of the phrase, had Lawrence's words— "don't give up the ship"—stitched onto a battle flag that he flew during the decisive Battle of Lake Erie in September 1813. Having to transfer from his badly damaged flagship at the height of the battle, he carried this flag with him. Aboard the U.S. brig *Niagara,* a victorious Commodore Perry penned a dispatch to General William Henry Harrison, which was to become famous in its own right: "We have met the enemy and they are ours." The venerable battle flag bearing Lawrence's immortal words is on display at the U.S. Naval Academy in Annapolis, Maryland. Although the original phrase "don't give up the ship" used to be strictly nautical, the expression has crept into common English usage as an exhortation not to give up. Commodore Perry's own famous words reporting his victory on Lake Erie achieved a life of their own when paraphrased some years ago by Walt Kelly's irrepressible cartoon character Pogo, who said, "We have met the enemy and he is us!"

Down the Hatch . . . (a drinking toast)

Cargo being lowered down a hatch into a ship's hold inspired the well-known toast,
"Down the hatch," which celebrates the act of drinking. The expression
is thought to date from the early 1930s, and has been attributed to author
P. G. Wodehouse.

Drogue Chute . . . (device used to slow down a vehicle)

When a space shuttle touches down after a trip into outer space, its considerable forward
momentum is slowed by a drogue chute that billows open behind it. But
the concept of a drogue chute is not a product of space-age technology.
From the Middle Low German word *dragge,* which means a drag anchor or
grapnel, the term *drogue* was originally used to describe a conically shaped
bag designed to slow a ship's drift to leeward, and to keep her headed into
the oncoming waves. In 1578, the redoubtable Elizabethan sea dog Francis
Drake used a drogue improvised from wineskins to dupe the Spanish into
thinking that his own ship, the *Golden Hind,* was slower than she really
was. This ruse led to Drake's legendary encounter with the *Nuestra Señora
de la Concepcion,* the great treasure ship better known by her nickname, the
Cacafuego (Shitfire). Under cover of darkness, Drake cut away the drogue,
then quickly overtook,
boarded, and seized the
fabulously rich Spanish
ship. Drake and the
Golden Hind returned
home after circumnavi-
gating the globe—as
triumphantly as a space
shuttle, but with consid-
erably less media coverage.

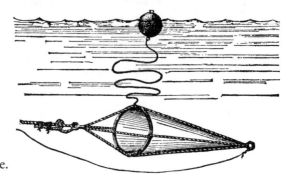

Dutchman's Breeches . . . (small patch of blue in an otherwise cloudy sky)

To a sailor, a patch of blue sky signifies the breaking up of a storm at sea. Even the smallest glimpse is viewed with optimism and said to be "enough to patch a Dutchman's breeches," the Dutch being famous for their thriftiness. In *The Sailor's Wordbook,* Admiral William Smyth described this portent of better weather as being "enough to make a pair of breeches for a Dutchman." The expression washed ashore where it is still used to describe a small patch of blue sky. Occasionally, less optimistic landlubbers use the expression in a variant form: "not enough blue to patch a Dutchman's breeches."

Dutch Courage . . . (bravery induced by a stiff drink, or the drink itself)

In the seventeenth century, long-standing trade rivalries and naval jealousies between England and the Netherlands erupted into the Anglo-Dutch Wars. The conflicts gave rise to a plethora of derisive "Dutchman" expressions. *Dutch courage* can be traced to the practice of certain Dutch admirals who allowed their men to drink a hearty libation of "square-faced gin" before engaging the enemy. The English took the opportunity to spread the story that their naval adversaries' courage was directly proportional to the amount of schnapps that they consumed. Thus, in 1873, Herbert Spencer noted in his work "Sociological Studies," "A dose of brandy, by stimulating the circulation, produces Dutch courage."

Ebb and Flow . . . (diametrically opposed changes)

The rhythmical, alternating, backward and forward movement of tidal current is known as its ebb and flow. Just as the ocean undergoes continuous tidal surges and recessions, so human beings experience a similar ebb and flow in the emotions and circumstances of life and love. For centuries, the ramifications of change, even though predictable, have proven fodder for literature. As Percy Bysshe Shelley wrote in his poem "Time":

> *Unfathomable Sea! whose waves are years!*
> *Ocean of time, whose waters of deep woe*
> *Are brackish with the salt of human tears!*
> *Thou shoreless flood which in thy ebb and flow*
> *Claspest the limits of mortality. . . .*

Fagged Out . . . (tired and at loose ends)

Rope that is used aboard ship tends to unravel and fray at the ends if it is not properly maintained. A frayed end is called the *fag end* and the rope itself is said to be "fagged out." The nautical etymology of the word *fag* is obscure, but it is thought to be a corruption of the word *fatigue*. Ashore, a person is said to be "fagged out" when exhausted or, in a figurative sense, beginning to unravel. The colloquialism "fag end" refers to the last part or remnants of something. The expression is used in this context by Laurence Sterne in his 1765 classic *Tristram Shandy:* "to be wove into the fag end of the eighth volume."

Fairway . . . (open path)

According to William Falconer in his 1789 *Universal Dictionary of the Marine*, a
fairway is "the path or channel of a narrow river, bay, or haven in which
ships usually advance in their passage up or down." Today, the game of
golf shares the word *fairway* with the nautical definition—on a golf course,
a fairway is the clear or open path, that is, the part of the course without
tees, putting greens, and hazards that provides a fair lie for the ball.

Fathom . . . (get to the bottom of things)

From the Anglo-Saxon *faethm* (to embrace), a fathom represents the span between the
two outstretched arms of a person of average size—approximately 6 feet.
Admiral William Smyth remarked that it was once defined by an act of
English Parliament as "the length of a man's arms around the object of his
affections." In *King Henry IV, Part I,* Shakespeare referred to the fathom,
known universally as a measurement for sea depths:

> *By heaven, methinks it were an easy leap,*
> *To pluck bright honour from the pale-face'd moon,*
> *Or dive into the bottom of the deep,*
> *Where the fathom-line could never touch the ground. . . .*

Nautical fathoms are gradually being replaced by soundings expressed in
metric values. The length of a man's arms around the object of his affection
is now equal to 1.8256 meters. Figuratively, by sea or by land, *fathom*
describes the process of delving deeply into an idea or concept in order to
get to the bottom of it. Harold J. Morowitz uses the metaphor
in his poem *To the Humpback Whales:*

> *We once had a philosopher named Melville*
> *Who maintained that you fathomed*
> *the secrets of the universe,*
> *But his name I only whisper to you, whales.*

Fend Off . . . (to push away)

Fending off is the process of pushing a vessel away, by spar, boathook, or fender, to prevent violent contact when coming alongside a dock, pier, or another vessel. *Fend off* and *stave off* are synonymous nautical terms. Ashore, both terms mean to "ward off," to "parry," or to "offer resistance" to something or someone. Thomas Jefferson used the expression in its colloquial form in his personal writings: "with fendings and provings of personal slanders."

Figurehead . . . (ornament)

Figureheads are carved, painted busts fixed just beneath the bowsprit of a vessel. They are highly ornamental and valuable works of art that reflect some aspect of the ship's personal identity, or that of the company or owner she represents. Beautiful as they are, figureheads are without practical function and ships could sail with equal mechanical efficiency without them. The only practical function that can be ascribed to them is in inspiring the crew. On land, *figurehead* describes a nominal leader whose good name and reputation lend credibility to a group or organization and, like a ship's figurehead, is purely ornamental. The practice was noted in an 1883 issue of *The Congregationalist:* "a mere diocesan figurehead with no opinions at all."

Filibuster . . . (delaying tactic)

Derived from the Dutch word *vrijbuiter* (freebooter or pirate), the English word *filibuster* was once synonymous with "buccaneer" or "pirate." During the eighteenth and nineteenth centuries, *filibuster* was used in the United States to describe gun runners, pirates, and an assortment of brigands, all of whom conducted raids, unauthorized military expeditions, and blockades off the coasts of the West Indies and Central America. After a politician described an adversary's stonewalling tactics as "filibustering against the

United States," *filibuster* came to describe obstruction of legislation by irregular maneuvering such as nonstop speech making. A headline in the *Boston Journal,* dated 20 February 1885, read: "Ex-Confederates Filibuster to Prevent Vote on Bill."

First-Rate . . . (excellent)

The mighty warships of the Royal Navy were once rated on a scale from one to six based on their size and the weight of ordnance they carried. Admiral Nelson's 2,163-ton flagship H.M.S. *Victory* mounted 100 heavy guns and was ranked as a first-rate ship of the line. The naval rating system was assimilated into everyday usage to describe degrees of excellence in general terms. In his 1749 novel *Tom Jones,* Henry Fielding wrote, "his natural parts were not of the first rate."

Fish or Cut Bait . . . (a command to act)

Fish or cut bait is a command to "get on with the job" or to "get out of the way" for someone who will. The expression, which became popular in the chaotic, postwar political arena of the 1870s, has been attributed to Republican Congressman Joseph G. Cannon who used the term during a rally. It is not known whether this expression was one that was popular at the time, or whether the political firebrand invented it during the heat of the moment. *Fish or cut bait* is a no-nonsense, direct command, but not quite as colorful as others attributed to Cannon, who was known to his congressional colleagues as "Foul-mouthed Joe."

Fits the Bill . . . (just right)

A bill of lading was a ship's manifest that itemized all goods being transported. When the cargo was unloaded from the hold, it was always carefully checked to see that the goods fit the bill by matching item for item. In a colloquial sense, something "fits the bill" when it is just right for its intended purpose.

Flake Out . . . (to drop out)

Flaking is the procedure of laying out the anchor chain on the deck of a ship for inspection. The length of the chain is ranged up and down the deck so that weak links can be located and replaced. In everyday shore parlance, *flaking out* means making a hasty withdrawal from an activity because of fatigue. Perhaps this unusual expression derives from the flaking out, or laying out, of a weakened anchor chain.

Flogging a Dead Horse . . . (an exercise in futility)

The band of variable calm in the Atlantic Ocean—roughly in the area of the Canary Islands—is known as the horse latitudes. They take their name from the Spanish *Golfo de las Yeguas* (Gulf of the Mares). It is thought that the Spanish name stems from a comparison between the unpredictable nature of the high-strung Arabian mare and the capricious nature of the wind in the area. In the days of sail, when a sailor signed up for the duration of a voyage, it was customary to pay him one month's wages in advance—but a sailor's money never lasted long in rollicking port towns. Once their advanced wages had been spent and the ship had put to sea, sailors felt as though they were working for nothing. Because it took approximately one month to reach the horse latitudes from most ports in England, sailors began the tradition of calling that first month at sea the "dead horse month." To mark its end, the crew celebrated by stuffing a canvas likeness of a horse with straw and marching it around the deck with great pomp and ceremony. The symbolic representation of the "dead horse" was then hauled aloft to the yardarm and cut adrift into the sea, as the sailors chanted, "Old Man [Captain], your horse must die!" Admiral William Smyth suggested that flogging a dead animal into activity was as much an exercise in futility as trying to get a wholehearted work commitment out of the ship's company while they were working off the dead horse month.

Flotsam and Jetsam . . . (odds and ends)

From the Latin *fluere* (to float), and *jacere* (to throw out), *flotsam* and *jetsam* are legal marine terms describing goods lost overboard with specific salvage rights applicable to each. *Flotsam* are goods that have been swept overboard and are found floating on the surface of the sea. *Jetsam* are goods deliberately thrown overboard or jettisoned to lighten the ship during an emergency. The metaphor *flotsam and jetsam* is used ashore to describe cast-off elements of society, or as a synonym for odds and ends. In the 4 December 1988 issue of the *Washington Post Book World,* Jonathan Yardley noted "three books that had escaped my attention; discovering them in the midst of 1988's flotsam and jetsam was a pleasure."

Fluke . . . (chance happening)

The origin of the word *fluke* is obscure. Some etymologists suggest that a possible source appears in the dialect of Northern England, where the word is still widely used. In any event, the word was known to nineteenth-century philosopher and economist John Stuart Mill who wrote that "The transfer of power has gone on by flukes and leaps in the dark." The nautical term *fluky* describes the wind at sea when it is light and variable; that is, wind that is not settled down enough to blow steadily from any one direction.

Fly-by-Night . . . *(dubious reputation)*

An accessory squaresail normally set on a temporary yard, the "fly-by-night" takes its name from the fact that it was easy to handle, making it particularly useful for sailing in the dark. Like the sail of the same name, the human "fly-by-night" is prone to nocturnal excursions and is "here today, gone tomorrow," frequently defrauding en route. In his 1822 novel *Maid Marian,* English author Thomas Love Peacock took a rather whimsical potshot at Robin Hood: "Would you have her married to a wild fly-by-night that accident made an earl and nature a deer stealer?"

Fogey . . . *(behind the times, old-fashioned)*

Francis Grose's 1785 *Dictionary of the Vulgar Tongue* defines *old fogey* as a "nickname for an invalid soldier." In the 1867 edition of *The Sailor's Wordbook,* Admiral William Smyth defined a *fogey* as an "old fashioned person, an invalid soldier or sailor." *Fogey* probably has its origins in the Scottish word *feggy* (or *fuggy*), which means covered with moss or grass.

Foul Up . . . *(confusion reigns supreme)*

From the Old English *ful* (grossly offensive to the senses), *foul* is an adjective with many nautical meanings, all indicative of a difficult or negative situation. An anchor is said to be "fouled" when it is entangled in its own cable or some other impediment. Sailors regard a fouled anchor with loathing but ironically, a fouled anchor is an insignia of both the United States Navy and Royal Navy. Its adoption as an official insignia dates from the waning years of the sixteenth century when a fouled anchor was incorporated into the arms of Lord Howard of Effingham, High Admiral of England. By land or by sea, a hopeless mess or entanglement is described as a "foul up." The use of this expression is so popular that it has taken the form of acronyms, the best-known being SNAFU ("situation normal, all fouled up") and, more recently, FUBAR ("fouled-up beyond all recognition"); FUBB ("fouled-up beyond belief");

and FUMTU ("fouled-up more than usual"). *Foul,* of course, was an "acceptable" substitute for a word not openly used in Victorian society or "polite company."

Founder . . . (to sink or fail)

The word *founder* is derived from the Latin *fundus* (bottom). A ship is said to "founder" when she takes on enough water to flood her hull and she sinks. To founder on land, or even in a figurative sense, is to fail completely. In Shakespeare's play *King Henry VIII,* the Lord Chamberlain speaks of the king's machinations in securing a divorce from his first wife: "But in this point all his tricks founder . . ." The word *founder* is frequently used interchangeably and incorrectly with the word *flounder,* which derives from a Dutch word meaning to struggle clumsily. Dramatist William Congreve used the word in his Restoration comedy *The Double Dealer:* "you will but flounder yourself a-weary."

Freshen the Hawse or Hawser . . . (to take a drink)

Although it is not clear which expression appeared first, both have nautical origins and date from the same period. *Hawse,* which stems from the Old Scandinavian word *hals* (neck or throat), is an opening in the ship's bow, through which the anchor cables pass. *Hawser,* a derivative of the Anglo-French *hauucor* (hoist), is a heavy rope or small cable used for a multiplicity of shipboard purposes. In nautical terms, to "freshen" is to "renew." Admiral William Smyth described the procedure in *The Sailor's Wordbook* in 1867: "to freshen the hawse (or the nip) is to pay out more cable so as to change the place of the exposed part to friction." Nineteenth-century sailors used the expression "to freshen the hawse" when describing the action of officers of the watch who, after a long spell on deck during cold weather, took a nip of spirits. In his 1827 novel *Red Rover* James Fenimore Cooper wrote, "Profiting by the occasion 'to freshen his nip,' as he quaintly called swallowing a pint of rum and water, he continued his narrative." Both expressions washed ashore where they are used interchangeably to mean "liven oneself up with a nip of spirits."

Fudge . . . (deceive)

Fudging is the act of deceiving by adjusting or making do in a careless, clumsy, or contrived manner. The origin of this expression is thought to stem from a mendacious real-life sea captain surnamed Fudge. Isaac D'Israeli, father of the great English statesman Benjamin Disraeli, related the following story in his 1791 collection of historical and literary anecdotes entitled *Curiosities of Literature:*

> *There was, sir, in our time one Captain Fudge, who upon his return from a voyage, how ill-fraught soever his ship, always brought home a good cargo of lies, so much that now aboard ship the sailors, when they hear a great lie told, cry out "You fudge it."*

As early as 1797, the expression had entered the public domain as a synonym for deceit or nonsense. As James Russell Lowell noted in *A Fable for Critics,* "There comes Poe with his Raven like Barnaby Rudge, / Three-fifths of him genius, and two-fifths sheer fudge."

Galoot . . . (a term of good-natured deprecation used to describe an ungainly or awkward person)

A military expression of unknown origin, *galoot* is defined in the 1812 *Dictionary of the Vulgar Tongue* by J. H. Vaux as "an awkward soldier." The word is used by Captain Frederick Marryat in his 1835 novel *Jacob Faithful:* "four greater galoots were never picked up." Smyth defined the term in his 1867 *Sailor's Wordbook* as "an awkward soldier or young, green marine." Galoots, like the lovable Gabby Hayes, run rampant in old black-and-white westerns.

Gangway! . . . (get out of the way!)

From the Old English *gangweg* (passageway), the noun *gangway* refers to a passage on a ship's upper deck or, more commonly, to the portable ramp that is used to get goods and people from a ship to a pier, or vice versa. Used as an interjection by sailors and landlubbers alike, *gangway* means "Get out of the way, I'm coming through!"

Gingerbread . . . (gaudy ornamentation)

During the Middle Ages, gingerbread cakes were fashioned into the shapes of people, animals, and letters of the alphabet and were then gilded. Later, during the heyday of the East India Company, the hulls of the company's merchant ships were decorated heavily with gilded scrollwork and ornate carvings. Because the ship's fancy work was reminiscent of the ornate medieval cakes, it became known as "gingerbread." The expression entered the public domain and is now synonymous with gaudy or tasteless ornamentation. Washington Irving used the word in his 1808 whimsical essay "Salmagundi, or the Whim-Whams and Opinions of Launcelot Langstaff, Esq. and Others": "two of those strapping heroes of the theatre who figure in the retinue of our gingerbread kings and queens."

Give a Wide Berth . . . (ample clearance)

"Berth" is a nautical term that indicates safe operating space for a vessel. To *give a ship wide berth* means to avoid her and keep a safe distance. The expression has similar application ashore, that is, not to approach something or someone too closely for fear of undesirable consequences.

Give Quarter . . . *(to show mercy or lenience)*

The origin of *give quarter* can be traced to the ancient practice of giving a captured knight or officer his life and liberty for ransom money, which was estimated at one-quarter of his yearly pay. Percy Bysshe Shelley used the expression in its metaphorical form in his 1818 work *The Revolt of Islam:* "There is no quarter given to Revenge, or Envy or Prejudice."

Grease the Ways or Skids . . . *(to smooth the way)*

On 31 May 1911, the world's largest and most luxurious ship majestically slid down the stocks at the Harland and Wolff Shipyard into Ireland's Belfast Lough. Twenty thousand pounds of tallow had been applied to the skids to facilitate a smooth launching. Having the skids greased does not, however, ensure smooth sailing, either figuratively or literally. The ship, the R.M.S. *Titanic,* was to make her apocalyptic maiden voyage the following year.

Great Guns! . . . *(nautical expletive)*

In *The Sailor's Wordbook,* Admiral William Smyth used this expression to describe "heavy cannons and officers of notable repute." Later, the term was applied to heavy weather, presumably because of the booming of the wind in storms at sea. In 1841, Charles Dickens noted such bluster in *Barnaby Rudge:* "It blows great guns, indeed. There'll be many a crash in the forest tonight." In everyday usage, "Great guns!" is an exclamation, roughly equivalent to "Holy smokes!"

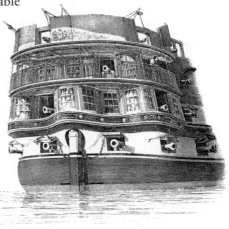

Grin and Bear It . . . (endure all with good nature)

The prolific naval lexicographer Admiral William Smyth gave the following definition of this popular phrase: "The stoical resignation to unavoidable hardship, which being heard aboard ship by Lord Byron produced the fine stanza in *Childe Harold* commencing 'existence might [sic] be borne.'"

> *Existence may be borne, and the deep root*
> *Of life and sufferance make its firm abode*
> *In bare and desolated bosoms; mute*
> *The camel labours with the heaviest load,*
> *And the wolf dies in silence not bestow'd*
> *In vain should such example be; if they,*
> *Things of ignoble or savage mood,*
> *Endure and shrink not, we of nobler clay*
> *May temper it to bear, it is but for a day.*

Grog . . . Groggy . . .
(spirituous drink, and the dazed, unsteady condition that it produces)

Admiral Sir Edward Vernon (1684–1757) was known throughout the Royal Navy as
"Old Grog" because he always appeared on deck in a dramatic cloak
fashioned from *grogram* (grosgrain)—a coarse fabric of silk, mohair, and
wool. Vernon was a flamboyant man of intemperate speech, a trait that did
not endear him to the office of the Admiralty. He also favored strict discipline
and temperance with regard to drink, a trait that did not endear him to his
men. In order to curb the incidence of drunkenness aboard ship, "Old Grog"
ordered that the daily rum ration of one pint neat issued to sailors of the fleet
was thenceforth to be diluted with water. Outraged sailors promptly named
the watered-down concoction *grog* after the none-too-popular admiral. Daily
rum rations were discontinued in the Royal Navy in 1970, but the popular
tradition lives on in song and verse. In his poem about sailor Sam Swipes,
Frederick Marryat, an early-nineteenth-century naval officer and writer, could
not resist the urge to comment on grog and its effects:

> *Sam Swipes, he was a seaman true,*
> *As brave and bold a tar*
> *As e'er was dressed in navy blue*
> *On board a man-of-war.*

> *One fault he had—on sea or land*
> *He was a thirsty dog;*
> *For Sammy never could withstand*
> *A glass or so of grog.*

> *He always liked to be at sea,*
> *For e'en on shore, the rover*
> *If not as drunk as he could be,*
> *Was always "half seas over."*

The gunner, who was apt to scoff
With jokes most aptly timed,
Said Sam might any day go off,
'Cause he was always "primed."

Sam didn't want a feeling heart,
Though never seen to cry;
Yet tears were always on the start,
The drop was in his eye.

At fighting Sam was never shy,
A most undoubted merit;
His courage never failed, and why?
He was so full of spirit.

In action he had lost an eye,
But that gave him no trouble;
Quoth Sam, "I have no cause to sigh,
I'm always 'seeing double.'"

A shot from an unlucky gun
Put Sam on timber pegs;
It didn't signify to one
Who ne'er could keep his legs.

One night he filled a pail with grog,
Determined he would suck it;
He drained it dry, the thirsty dog!
Hiccupped, and "kicked the bucket."

Grog Blossom . . . (red nose)

This colorful term is the sailor's description for the bulbous red nose frequently seen on grog aficionados such as Sam Swipes. (See also "Grog . . . Groggy.")

Gunboat Diplomacy . . . (force)

The earliest gunboats were small, light-draft vessels fitted for carrying one large-caliber gun. Gunboats have changed dramatically over the centuries but the implication of their strategic use has not. *Gunboat diplomacy* is an allusion to influence by force of arms rather than by skilled negotiations. Its use is thought to date from the days of Teddy Roosevelt and the Spanish-American War.

Half Seas Over . . . (helplessly drunk)

The expression *half seas over*, meaning halfway across the sea, was seen in print as early as 1551. In 1618, Sir Walter Raleigh referred to "ships that ride it out at anchor half seas over between England and Ireland." The origin of the expression's use to describe an advanced state of inebriation is somewhat obscure. It is thought to stem from the Dutch *op zee ober* (overseas beer), a potent brew introduced into England from Holland in the seventeenth century. Since *op zee ober* sounded reasonably like "half seas over," at least to English ears, corruption of the Dutch was inevitable. In any event, the effects of this brew seem to be considerable, for it has given rise to a number of expressions, all implying some degree of inebriation.

Hand over Fist . . . (rapid progress)

In the heyday of wooden sailing vessels, the speed and
agility with which a sailor could climb aloft
into the rigging was highly valued both as a
technical skill and as a source of pride for
the individual sailor. The nautical expression
hand over hand originated with English
sailors as a literal description of the tech-
nique used in climbing up or down a rope,
or hauling in or letting out a sail. The idea
of hauling in a sail or making a rapid ascent
on ropes soon acquired the figurative meaning of continuous, rapid advancement. In
1813, Captain Frederick Marryat used
the expression in his novel *The King's Own:*
"The frigate was within a mile of the lugger
and coming up with him hand over hand."
Noting that these climbing and hauling techniques involved a free hand
passing over the other fist in which the rope was clenched, it is thought that
American sailors changed the expression to *hand over fist.* In the twentieth-
century context, *hand over fist* has become synonymous with financial gain,
the result of a rapid ascent up the ladder of success.

Hard and Fast . . . (rigid)

A ship that has been run aground or beached is said to be *hard and fast,* that is, hard
aground and fixed fast, or immovable. In a figurative sense, the expression
hard and fast refers to rules strictly laid down and rigidly enforced. Joseph
Warner Henley used the expression in this sense in a speech to the English
House of Commons in 1867, in which he questioned "whether the
franchise is to be limited by a hard and fast line."

Hard Up . . . (in a general state of need)

Hard up in a clinch and no knife to cut the seizing was the sailor's way of saying that he
was beset by misfortune with no way to cut himself free. *Hard up,* the
shortened colloquial expression that washed ashore, means a general state
of want or shortage of funds. Characters in Charles Dickens's *Bleak House*
use the expression: "He was in need of copying work to do and was . . .
hard up." A variant nautical source of this expression is the command *hard
up the helm!* This order was given in stormy seas when the tiller had to be
brought sharply to windward in order to turn the ship's bow away from the
wind. In a figurative sense, being "hard up for money" would certainly
involve weathering rough seas. Still another derivation can be seen in the
traditional phrase "helm hard up for Poverty Bay."

Haze . . . (humiliating horseplay)

From the Old French word *haser* (to affright, scare, or punish by blows), *hazing* was the
practice of making life as miserable as possible for crews aboard trading
ships during the age of sail. Hazing came in the form of bullying and
disagreeable busywork, which was doled out to the extent that crews were
deprived of legitimate hours of rest. In his classic tale *Two Years Before the
Mast*, Richard Henry Dana described how Captain Frank Thompson of the
brig *Pilgrim* hazed his crew. The sadistic captain assembled the crew on
deck in the rain and made them stand for hours in absolute silence, picking
oakum—tarred hemp fibers used for caulking seams aboard ship. This was
a tedious job generally reserved for old and infirm inhabitants of work-
houses. In addition to captains like Thompson who frequently asserted their
authority by hazing the crew, so-called "Bucko Mates" drove their victims
by brutality, making their lives aboard ship a living hell. Sailors could
become so exhausted as a result of hazing that they were unable to main-
tain their balance or footing while working aloft and, consequently, fell
to their deaths. In Robert Louis Stevenson's *Treasure Island*, a member of the

mutinous pirate crew roars a defiant, "I'll be hanged if I'll be hazed by you, John Silver!" In more recent years ashore, *hazing* has come to mean the initiation of a newcomer by rough, humiliating horseplay.

Head . . . (bathroom facilities)

The bows, or heads, of Roman galleys were once fitted with ornate bronze beaks that served as ramming instruments. As sailing ships developed, the "beakhead" was the name given to a structure projecting forward from the stem and the bowsprit. This was essentially a work platform, decked with grating and open to the sea below. The constant flushing action of the waves washing through the grating made the beakhead an ideal lavatory. While the nature of shipboard plumbing has changed considerably, the name has not. *Head* remains a popular reference to shipboard or land-based sanitary facilities for sailors and landlubbers alike.

Hell's Bells! . . . (nautical expletive)

According to old nautical dictionaries, the comparatively mild *Hell's Bells!* is a shortened version of the nineteenth-century expletive "Hell's Bells and Buckets of Blood!" Cursing and sailing are apparently compatible, because the expression "he swears like a sailor" is very much a part of English vernacular. In his poem *The Sailor to His Parrot*, William Henry Davies described how his fine-feathered friend makes an "endless song" of colorful curses learned aboard ship. Once ashore, the "foul-mouthed wretch" sings his naughty tune to horrified priests and unsuspecting widows.

Helm . . . (position of control)

The *helm* is the function, or sometimes the instrument (tiller or wheel), of steering a ship, thereby controlling her direction. Ashore, a person is at the helm if in a position of leadership or control. In his tragedy *Becket*, Alfred, Lord Tennyson wrote, "no forsworn Archbishop shall helm the Church."

High and Dry . . . (left in the lurch)

A beached ship, or one that is up on blocks and in the yard for repair or storage, is said to be "high and dry." Figuratively, a person is "high and dry" when left in the lurch or in an awkward position as a result of uncontrollable circumstances.

Hijack . . . (illegal seizure of goods in transit)

There are many suggestions for the origin of the word *hijack,* a nineteenth-century Americanism that originally meant the coercion or illegal seizure of a person. In all versions of the word's possible origins, the plot line and outcome are the same: one person is seized forcibly by another. The stories vary only in the identity of the person responsible for the seizure or the use of the key word, *hijack.* According to the nautical version, "Hi, Jack!" was a prostitute's come-on to a lonely sailor on shore leave. Anticipating a pleasurable encounter with the lady of the night, a distracted, unwary sailor was instead struck from behind and knocked senseless by one of "the lady's" confederates. He was then sold to a ship in need of a crew. The word *hijack,* also spelled *highjack,* now means the illegal seizure of cargo or a vehicle in transit, or sometimes even the theft of ideas.

Hodgepodge . . . (jumble)

This colorful expression has its origins in the culinary arts and in marine salvage law—quite a *hodgepodge* of disciplines. In Anglo-French, *hotchpot* was used to describe a dish made in a single pot in which many ingredients were mixed. During the thirteenth century, *hotchpot* entered the realm of legalese, as it was used to describe a process of property division. Its specific nautical application referred to the process of gathering property and cargo that had been damaged and strewn about because of a collision between two ships. Based on the premise that both ships contributed to the loss, the remaining goods were, by the process of *hotchpot,* divided equally between the two shipowners. In a metaphorical sense, *hotchpot* (corrupted to *hodgepodge*) is now synonymous with any confusion and disorder. As early as 1588, John Udall described "schisms that make a hotchpot of true religions."

Hooker . . . (prostitute)

The word *hooker* was seen in print in the United States as early as 1845, a fact that dispels the myth that General Joseph Hooker's Civil War camp followers gave rise to the word's slang usage. However, *hooker* continues to baffle etymologists. Like the word *hijack,* there are many and varied suggestions as to its possible origins. Of course, there is a nautical hat in the ring. A *hoecker,* or *hooker,* was a short, squat fishing rig favored by the Dutch during the seventeenth and eighteenth centuries. Some authorities believe that the slang use of the word *hooker* originated with the Hook of Holland—a spit of land where the ladies of the night would wait for sailors to come ashore. Sailors also used the word, either in a deprecating way or fondly, to describe any vessel that had lost the blush of youth, or that had aged none too gracefully, and had come down a peg or two in the maritime world, perhaps an apt description of the hooker's figurative counterpart. A December 1865 article in the *London Telegraph* described how sailors leaving their ship gave "the old hooker a hearty cheer." The slang expression for a prostitute may also simply be a broadened and more encompassing application of the word *hooker,* once defined as a thief, who literally used a hook to snatch away the belongings of his victim. Perhaps the ladies of the night did a bit of literal hooking on the side, and assisted the thieves as they plied their own trade. As Thomas Harman wrote in 1567, "These hookers be most perilous and wicked knaves."

Hotshot . . . (a show-off)

Iron cannon balls were sometimes heated in galley fires and carried in buckets to different parts of the ship to provide a bit of warmth on cold, damp nights at sea. According to a popular but incorrect belief, this practice gave rise to the colloquialism *hotshot.* Today, a *hotshot* is one who is skillful, but showy and aggressive—an allusion to a reckless, bold, hothead who shoots a firearm eagerly and enthusiastically. Thomas Middleton described the type in *Father Hubbard's Tales* in 1604: "To the wars I betook me and ranked myself amongst the desperate Hot-Shots."

Hulk . . . (big, clumsy)

This term comes from the Old French word *hulque,* a large, flat-bottomed transport vessel. The *hulc,* which originated in Northern Europe in the Middle Ages, was a large, shallow-draft cargo vessel. When no longer fit for sea duty, *hulcs* were frequently dis-mantled and converted for nonsailing uses—prison, storehouse, quarantine, or other functions that did not require them to move.

By extension, *hulk* describes the body of a large, obsolete, or abandoned ship. Similarly, ashore, the word is used to describe something (or someone) that is large, clumsy, and unwieldy. In a pungent rendition of the "I haven't seen you since you were a little boy" cliché, Henry Brooke wrote in 1767: "You are grown to a huge, hulking fellow since I saw you last" *(A Fool of Quality)*. Perhaps Brooke anticipated the modern American television series *The Incredible Hulk*.

Hunky-Dory . . . (enjoyable, pleasant, and "A-OK")

This expression is thought to have its origins in *Honkidori,* the name of a street in the Yokohama, Japan, waterfront district where a sailor on shore leave could find just about anything his heart desired.

Idler . . . (loafer)

From the Old English word *idel,* meaning empty or useless, the word *idler* refers to someone who has nothing to do. In 1534, Thomas Dorset used the word in his *Suppression of the Monasteries:* "I having nothing to do as an idler went to Lambeth to the bishop's place to see what new is." The word *idler* also had a specific nautical application. In his 1789 *Universal Dictionary of the Marine,* William Falconer defined an idler as a member of the ship's company who was subjected to constant duty during the day, and therefore was not required to keep a night watch except in case of an emergency. The hardworking carpenter, cook, sailmaker, and sickbay men were called idlers, but certainly were not idle, despite what the name seems to imply.

In the Same Boat . . . *(sharing risks)*

A boat is a small, open vessel that is at risk when in open waters. To avoid capsizing, everyone in the boat must work together in order to get safely ashore or back to the ship. The expression "in the same boat" is an allusion to the dangerous and sometimes lonely plight of mariners at sea in a boat. Ashore, people who experience the same difficulty, whether separately or collectively, are considered to be *in the same boat.*

Jack . . . *(man of the people)*

Jack or *Jack Tar* is a popular appellation for British sailors. From the nickname for John, "Jack" represents a man of the common people, a good-natured buddy to every person aboard ship. The expression "quick as you can say Jack Robinson" was noted by Admiral William Smyth in *The Sailor's Wordbook* in 1867. The British terms "Jack Ashore" or "Jack the Lad" denote a spirit of reckless abandon, much like a sailor on shore leave. In his 1915 novel *Victory,* Joseph Conrad remarked, "and it was in something of the Jack Ashore spirit that I dropped a five franc note in the gravy boat."

Jaunty . . . *(sprightly manner)*

The adjective *jaunty* is said to be an anglicized, phonetic representation of the French word *gentil,* meaning a sprightly, easy manner. Used as a noun, *jaunty,* or *jonty* as it was sometimes spelled, means the master-at-arms aboard a British warship—the officer responsible for enforcing all rules and regulations, as well as for meting out punishment. The nautical *jaunty* is said to be a corruption of *gendarme,* the French word for "police officer."

◆ ◆ ◆

Johnny-Come-Lately . . . (newcomer)

Applied either contemptuously or in a good-natured way, *Johnny Raw* was an appellation
for inexperienced British sailors. In 1888, Robert Louis Stevenson used the
expression in his book *Kidnapped:* "You took me for a country Johnny Raw
with no more mother wit or courage than a porridge stick." In America, the
saying evolved into *Johnny-come-lately,* and by land or by sea, the expression
is used to describe newcomers or novelties. Pop artists, The Eagles, can still
be heard singing about "a new kid in town."

Jonah . . . (one who brings bad luck)

In the classic sea adventure story *Captains Courageous,* Rudyard Kipling described a
Jonah as "anything that spoils the luck." The expression clearly originates
in the biblical story of Jonah. As a result of divine punishment, the
unfortunate Old Testament prophet suffered a series of seagoing
misadventures until his comrades threw him overboard for the bad
luck he had brought them, whereupon he was swallowed by a whale
and spewed on land three days later. The use of this expression is popular
ashore, where it describes someone who brings misfortune. Cartoonist
Al Capp caricatured the Jonah concept in his comic strip *Li'l Abner* by
always drawing a black cloud of doom and gloom over the head of his
character with the unpronounceable name of Joe Btfsplk.

Jury-Rigged . . . (makeshift)

In nautical parlance, *jury* denotes a temporary contrivance, such as a jury-mast or jury-
rudder used to get a ship under way after she has been disabled. The word
jury is derived from the Latin word *jurare* (to swear an oath), but the origin
of *jury* in its nautical application is not known. It has been suggested that
the word was coined by cynical sailors and is a shortened version of injury-
rigged. Anything temporary or improvised is considered to be "jury-rigged."
In his novel *The Adventures of Peregrine Pickle,* Tobias Smollett referred to a

sailor with a peg leg as a "jury-legged dog." The most whimsical use of the expression dates from 1666. In his narrative of the Dutch Wars entitled *Instructions to a Painter,* Sir John Denham warned, "Guard thy Posterior lest all be gone; Though Jury-Masts thou hast, Jury buttocks none." Frequently mispronounced as "jerry-rigged," the term is used ashore to describe something improvised or thrown together hastily.

Keelhaul . . . (rough reprimand)

Keelhauling was a barbaric form of punishment thought to have been devised by the Dutch and adopted by other navies during the fifteenth century. The unfortunate sailor subjected to such punishment was weighted down, hauled up to one yardarm, and dropped suddenly into the water far below. In *A Dialogical Discourse,* written in 1634, Captain Nathaniel Boteler noted that while the victim was underwater, a "great gun was discharged . . . which is done as well to astonish him so much the more with the thunder of the shot, as to give warning to all others of the fleet to look out and be wary of his harms." Stunned by the noise and the buffeting against the ship's hull, numb with cold, and nearly asphyxiated from being submerged, the sailor was then dragged under the ship's keel and hauled aloft to the opposite yardarm by means of ropes that had been prerigged to his body. The keelhauling process was repeated until enough punishment had been inflicted or until the victim was dead, whichever came first. An even more infamous form of punishment was known as *keelraking,* or dragging a sailor underwater the length of the ship. "Keelhauling" and "keelraking" gradually went out of fashion during the early part of the eighteenth century. But the word lives on ashore, where *keelhauling* has come to mean a severe tongue lashing or an exaggerated reprimand from a superior, or, as Admiral Smyth punned in his *Sailor's Wordbook,* "undergoing a great hardship."

Knock Off . . . *(to stop instantly)*

According to old nautical dictionaries, this expression was once a standard order aboard
sailing ships and is still commonly used at sea today. It stems from when
galleys used to be rowed to the rhythm of a mallet striking a wooden block.
When the knocking stopped, it was a signal to stop rowing and rest. In
Two Years Before the Mast, Richard Henry Dana used the expression in its
colloquial form: "after we had knocked off work and cleaned up the decks
for the night."

Know the Ropes . . . *(skill and experience)*

The rigging in a square-rigged ship was a vast, complex network of cordage. It included,
among other things, the ropes used to support the yards and masts, as well
as those used to hoist, lower, and trim the sails. This complicated system
contained hundreds of separate pieces, each having a name and particular
function. Running aloft as well as fore and aft, the end of each rope was
secured to a belaying pin and identified by its position on rails running the
length of each side of the ship. The mastery of this crucial and complex
system separated old salts from "Johnny-Raws"—the men from the boys.
It was considered so important that discharge papers were once marked,
"knows the ropes," thus constituting an honorable discharge.

Laid Up . . . *(temporarily disabled)*

By definition, a ship that is *laid up* is one that is dismantled, disabled, or moored, either
for want of employment or because she is unfit for sea duty. Ashore, a
person is said to be "laid up" when ill or disabled, and not fit to carry out
daily activities.

Landlubber . . . (opposite of a sailor)

From the Old English word *londloper,* meaning one who runs up and down the land, the word *landloper* was originally applied to a vagabond. In his *Book of Martyrs* published in the 1580s, John Fox paraphrased a papal bull of the previous century: "certain arch heretics have risen and sprung up being landlopers, schismatics and seditious persons." The word *loper* eventually became corrupted to "lubber," and "landlubber" became the mariner's contemptuous expression for landlubbers and freshwater sailors alike. In 1875, F. Burton commented, "The philosophic landlubber often wonders at the eternal restlessness of his naval brother-man."

Landmark . . . (point of reference)

A landmark is a prominent, fixed object on the landscape such as a church tower, lighthouse, or mountain whose position is marked on a navigational chart. It enables a navigator to establish the ship's bearings. A *landmark decision* is

a distinguishing fact or event that serves as a guide or boundary in similar decision making. In his work *Utilitarianism,* the English philosopher John Stuart Mill used the word metaphorically: "This man . . . whose system of thought will long remain one of the landmarks in the history of philosophic speculation . . . "

Latitude . . . (freedom to speak or act without constraint)

From the Latin *latitudo* (breadth), *latitude* represents a vessel's angular position north or south of the equator. Ashore, *latitude* describes a range within which one has the freedom to act at one's own discretion, unhampered by constraint or by narrow restrictions. In 1749, Henry Fielding used the word in its figurative sense in his novel *Tom Jones:* "He gave a latitude towards his friend's tongue and desired him to speak plainly what he knew."

Launch . . . (to enter boldly, or to start a new course of action)

From the Old French *lancier,* meaning to pierce or throw, a ship is "launched" when she is sent down the ramps from her building site on land, "pierces" the water, and is set afloat. Colloquially, the word means to enter boldly, as Samuel Richardson did in *Clarissa* (1748) when he "launched a guess." The word also means starting upon a different course of action, much as John Bunyan did in *Pilgrim's Progress* when he was "launched again into the gulf of misery."

Lay of the Land . . . (prominent features)

The word *lay* is one used frequently by sailors. It has many nautical meanings and applications. When a ship makes landfall, mariners study how the land lays along the horizon in an attempt to identify, among other things, familiar or prominent features and prospects for anchorage. In a metaphorical sense, one studies the lay of the land by determining the salient features and characteristics of a situation, along with its risks and benefits prior to becoming involved in or making a commitment to it.

Leeway . . . (freedom of action)

Leeway is the lateral drift of a vessel, or the distance she is forced to leeward of her intended course by the sideways pressure of the wind. Ashore, one is given leeway when allowed greater freedom of action and the space or time necessary to catch up. During his pilgrimage to the western frontier in 1835, Washington Irving described a steer as it "made great leeway toward the corn-crib filled with golden ears of maize."

Lifeline . . . (something or someone on whom life depends)

Lifelines are ropes or wires rigged fore and aft along the deck of a ship. They aid in saving crew from being washed overboard in heavy seas by providing a secure handhold. Old salts alluded to a mate who had died by saying that he had "unrove [unraveled] his lifeline." The word *lifeline* is frequently used ashore to mean anything or anyone, in a figurative or literal sense, on whom life depends. In 1884, Edmund Ufford Smith penned a Revivalist hymn around this nautical metaphor:

> *Throw out the lifeline across the dark wave,*
> *There is a brother whom someone should save,*
> *Throw out the lifeline, throw out the lifeline,*
> *Someone is sinking today.*

Limey . . . (British cousin)

Lime-juicer, or *limey,* was originally used by American sailors when referring to their British counterparts. The word had its origins in the Royal Navy regulation that called for a daily ration of lime juice to be issued to sailors to prevent scurvy. Formerly an appellation that carried a slight degree of contempt, *limey* has mellowed considerably and is now used by Americans and others as a term of affectionate regard for Brits in general.

Loaded to the Guards or Gunn'ls . . . (intoxicated)

A person may "drink like a fish," but if he is intoxicated he is *loaded to the guards* or *to the gunn'ls* and not to the gills. Plimsoll lines—named for the Victorian social reformer Samuel Plimsoll—are markings painted on the sides of vessels to indicate safe loading levels. Sailors of the period often referred to the markings as "guards." Anyone contemplating getting "loaded" to his guards should heed the words of the well-known song *The Unseaworthy Ship:* "Honor to Plimsoll his labor will save / Thousands of brave men from watery graves." The other "loaded" analogy is to the *gunwale* (pronounced "gunn'l"), which is the heavy plank at the top of a ship's side. A ship loaded to the gunn'ls would be loaded indeed!

Loggerheads . . . (stubborn, head-to-head conflict)

There are two definitions of a shipboard *loggerhead,* the first being a squat, wooden post around which a hand-launched harpoon line was controlled after a whale had been struck. The second nautical *loggerhead* was a tool of a different nature—a solid ball of iron the size of a person's fist, to which a long handle was attached. The ball was heated in the galley fire, then plunged into a bucket of pitch in order to bring the pitch to a consistency that could be poured into shipboard seams in need of caulking. This loggerhead became a lethal head basher when flung at an enemy or used as a weapon in hand-to-hand combat at sea. The inference of thickheadedness or uncompromising stupidity is inherent in the metaphor, and Shakespeare used *loggerhead* in this context frequently, as in *The Taming of the Shrew:* "you loggerheaded and unpolished grooms."

Look One Way and Row Another . . .
(nautical equivalent of a hidden agenda)

The analogy inherent in this metaphor refers to a rower, who sits facing one way while moving the boat in the opposite direction. The expression was known to the Earl of Aylesbury who published his memoirs in 1728: "certain it was that in her [the Queen's] court there were persons who looked one way and rowed another." Anyone who "looks one way and rows another" should heed the words of Thucydides. In the fourth century B.C., this eminent Greek naval historian wrote words to the effect that a collision at sea can ruin the entire day! The concept of oars seems to be ripe for metaphorical picking. "To rest on one's oars" means to suspend efforts; to "have an oar in everyman's boat" describes having a hand or meddling in everyone's affairs. In a happier mode, to "pull at one's oar" means to do one's share in a cooperative effort.

Loophole . . . (a way out)

A nautical *loophole* was a small aperture in the bulkheads and other parts of a merchant ship through which small arms could be fired at an enemy trying to board her. Metaphorically, a *loophole* is an ambiguity in the law that creates an opportunity for escaping the true legal intent. The expression and the practice of finding such convenient ambiguities was even noted by our founding fathers. In 1807, Thomas Jefferson commented, "what loophole they will find in the case when it comes to trial, we cannot foresee."

Loose Cannon . . . (recklessly and dangerously out of control)

When cannons on a wooden warship broke loose from their restraining tackle, they posed a serious threat to life and limb as they crashed into personnel and through bulkheads. In 1545, as Henry VIII's great ship the *Mary Rose* sailed out of Portsmouth harbor to engage the French fleet, she was flooded through her lower gunports. When she began to list sharply to starboard and her twenty heavy guns began to break free from their carriages, the fate of the *Mary Rose* was sealed. As an incredulous King Henry watched from a nearby promontory, the pride of his fleet sank quickly to the bottom with a loss of nearly all hands. Loose cannons have contributed to many disasters at sea, but not all loose cannons are found aboard ship. Some are found in the political arena, as Maureen Dowd reported in the 10 September 1989 issue of *The New York Times:* "he seemed to fulfill predictions by White House advisors that he would be a black sheep, a bull in a china shop, . . . a loose cannon. . . . A Mr. Aggressive Steam Roller."

Lopsided . . . (unbalanced)

In a figurative or literal sense, anything considered to be disproportionate or lacking in balance and symmetry is referred to as "lopsided." In his 1789 *Universal Dictionary of the Marine,* William Falconer defined *lopsided* (sometimes written "lap-sided") as "the state of a ship, which is built in such a way as to have one side heavier than the other; and by consequence, to retain a constant heel or inclination toward the heavier side." *Lop* also describes a sea condition when the waves are short and choppy. Sailing on a lop sea is far from a smooth ride, as the following note from an 1847 issue of *The Illustrated London News* indicates: "there being a lop on, the boat lurched to the windward."

Lower the Boom . . .
(to reprimand harshly, to stop someone from doing something)

A boom is a long spar or pole used to extend the bottom of certain sails; or, it can be a spar that extends upward at an angle from the foot of a mast from which there are suspended objects to be lifted. Derrick, the famous hangman during the reign of Queen Elizabeth I, devised the prototype for the ship's boom—a hoist that still bears the inventor's name. Ashore, *lowering the boom* on someone means to call that person harshly to account. This can be done severely enough to leave one's ears ringing, as if hit by a real boom.

Maelstrom . . . (turmoil)

This famous whirlpool off the west coast of Norway reputedly sucked in and destroyed all vessels that came within a wide radius of its vortex. *Maelstrom* takes its name from the Early Modern Dutch *malen* (to whirl around), and from *strom* (stream). The Elizabethan chronicler Richard Hakluyt noted "a whirlpoole called the mailstrome . . . which maketh such a terrible noise that it shaketh the rings in the doors of the houses of the said islands." In a metaphorical sense, a *maelstrom* represents a situation or an emotion that is

turbulent or violent. In 1855, author J. Abbott described the horrors of the
Napoleonic Wars: "an accumulated mass, in one wild maelstrom of
affrighted men, struggling in frantic eddies."

Mainstay . . . (chief support)

Taking its name from the chief shipboard mast that it supports, *mainstay* is a nautical
term that has come ashore where it also means "chief support." Even
President Thomas Jefferson was familiar with the term, as he showed in
referring to "the points of contact and connections with this country which
I consider as our surest mainstay under any event."

Making Both Ends Meet . . .
(to make do with what is at hand, no matter how meager)

The origin of this expression, frequently attributed to naval surgeon and novelist Tobias
Smollett, refers to the once common practice of splicing the ends of rope
together in order to cut shipboard operating expenses. In 1748, Smollett
used the expression metaphorically in *The Adventures of Roderick Random:*
"He made shift to make the two ends of the year meet." The figurative
usage, however, predates Smollett by a century. William Beveridge used the
phrase in 1662 in his *Sermons:* "worldly wealth he cared not for, desiring
only to make both ends meet."

Making Headway . . . (make progress)

Headway is defined as a ship's forward movement through the
water. Ashore, the term denotes forward
momentum of a general nature, as Jessop noted in
1887: "rarely do the demagogues make headway."

Mind Your P's and Q's . . . (mind your manners)

There are two plausible nautical derivations for this expression. It is said that wives admonished their seafaring husbands about soiling their peacoats (p's) with their tarred, greasy pigtails or queues (q's). The other version derives from the credit often extended to sailors by seaport pubs. A running total, reckoned in the number of pints and quarts consumed, was kept on each sailor. The petty officer, who was responsible for having a full and sober complement aboard ship, would remind his rambunctious charges during the course of an evening's carouse to *mind their p's and q's* (pints and quarts). Some authorities, however, suggest that the expression originated with children learning the alphabet. Since the letters *p* and *q* both have tails, it was difficult for the fledglings to tell them apart. They were told to "mind p's and q's." The nautical origins seem more credible because in modern usage, the expression is still a mild command to mind one's manners.

Miss the Boat . . . (lost opportunity)

Sailors on shore leave who had not minded their *p*'s and *q*'s (see preceding entry) were sometimes so groggy or half seas over the next morning that they literally missed the boat and did not make their intended voyage. They are not the only ones who can be left stranded on a dock, however. A legendary incident of "missing the boat" took place on stage at the old Metropolitan Opera House during a performance of *Loenghrin*. A swan that did double duty as a boat appeared on cue to collect Loenghrin for his exit. Because of a technical glitch, however, the swan-boat slipped off its track and departed before heldentenor Leo Slezak could climb aboard. Undaunted by the missed opportunity, Slezak turned to the audience and asked, "Wann geht der nächste Schwann?" (What time does the next swan go?)

◆ ◆ ◆

Monkey . . .
(a diminutive for a child, or, as a verb, to play foolish or mischievous tricks)

Monkey Pump . . .
(a straw for sucking up liquid, also synonymous with drinking from a bottle)

In the Middle Low German version of *Reynard the Fox* (1498), *Moneke* was the name of
Martin the Ape's son. Subsequently, *Monkey* came into popular English
usage as a term of playful contempt for young people. In *Don Juan,* Lord
Byron refers to a "little, curly-headed good-for-nothing, And a mischief
making monkey from birth." Nautically, anything diminutive is called a
monkey, whether wooden casks, blocks, or engines. A type of small coastal
trading vessel that was popular in the sixteenth century was called a monkey.
The little boys who assisted the gunners aboard warships were "powder
monkeys." A "monkey jacket" was a thick, close-fitting coat of blue serge,
cut short in order to free a sailor's legs for climbing aloft. Most popular
with sailors was the "monkey pump"—a straw or quill inserted through a
hole bored in a cask of wine or spirits. During the War of American
Independence, British sailors in the Caribbean drained the milk from
coconuts and refilled the shells with rum, which they sucked out at their
leisure aboard ship. Although sipping coconut milk seemed a harmless
enough activity, it took the Admiralty a long time to figure out the source
of an alarming rise in shipboard drunkenness. The practice, called *sucking
the monkey,* is now synonymous with drinking from a bottle. In *Beggar Girl*
(1797), Mrs. A. M. Bennett noted, "she is a goodish wench in the main, if
one keeps a sharp look-out after her, else she will sup [suck] the monkey."

Nausea . . . (miserable sensation in the pit of the stomach)

The word *nausea* springs directly from the Greek word *naus* (ship), a vehicle that causes the condition in one of its most virulent forms. In a figurative sense, *nausea* describes a feeling of strong disgust or loathing. As John Spencer wrote in his 1663 work *Prodigies:* "that nausea which tedious repetition creates in the soul of men."

Nautical . . .
(the art of seamanship and navigation; from the Latin nauticus, "a sailor")

Sixteenth-century astronomer Johannes Kepler spent his life attempting to fathom the motion of planets floating in the cosmic ocean of space. In *Somnium (The Dream),* one of the first works of science fiction, Kepler imagined being borne to the moon by "celestial ships with sails adapted to the winds of heaven." His "dream" has become reality—a new age of sail and exploration has dawned. Twentieth-century humans are no longer constrained to sail only earth's seas. A new breed of sailors looks to the heavens as astronauts (star-sailors) and cosmonauts (universe-sailors) navigate Keplerian trajectories into the boundless sea of space. The race to be first in space has given way to international cooperation, and imagined conflicts are kept at bay by the fictional starship *Enterprise.*

Nippers and Half-Pints . . . (general references to small children)

A vital part of weighing anchor, particularly in a big man-o-war, was to bind, or nip, the heavy anchor cable to a lighter line, called the messenger, which ran around the capstan (winch). The lads who worked the nipping lines were called *nippers.* From the sixteenth to the eighteenth centuries, agile young boys were trained by cutpurses and pickpockets to assist them in their trade. Because of their size, the boys were known as "nippers," possibly an allusion to the "nipperkin," a measure of a half-pint or less that was used for spirits. Gradually, the term came to include all little boys who were legitimately employed as "gofers" to carters, street vendors, and the like.

No Man's Land . . . (neither here nor there)

Generally accepted as dating from World War I, the expression *no man's land* has much
earlier nautical origins. In his 1789 *Dictionary of the Marine,* William
Falconer stated that the expression *no man's land* "derives from a situation
of being neither on the starboard or larboard [old term for port] side
of the ship, nor in the waist or forecastle but being situated in the middle
part of both places." The expression was known to Thomas Bailey Aldrich
in 1877 when he used the metaphor in his short poem "Identity":

> *Somewhere — in desolate windswept space —*
> *In Twilight land — in No man's land —*
> *Two hurrying Shapes met face to face,*
> *And bade each other stand.*
> *"And who are you?" cried one agape,*
> *Shuddering in the gloaming light.*
> *"I know not," said the second Shape,*
> *"I only died last night."*

No Room to Swing a Cat . . . (cramped space)

Letting the Cat Out of the Bag . . . ("spilling the beans")

Some authorities believe that this expression derives from the fact that the cramped
confines between decks in old sailing vessels afforded very little room for a
sailor to swing his hammock or cot, a word easily corrupted to "cat." Still
others believe that the metaphor springs from a sturdy, but comparatively
small, sailing collier known as a "cat," once popular in northern Europe.
The expression *no room to swing a cat* may refer to the inability of the cat to
swing at anchor within the close quarters of a small port. The most popular
explanation for this expression, however, originates with that infamous
instrument of punishment, the "cat-o'-nine-tails"—a whip consisting of
nine pieces of cord approximately 18 inches long with three knots tied in

each cord, affixed to a short, stiff length of rope used as a handle. The nearly 2-foot cat, plus the length of the flogger's outstretched arm, required ample clearance, hence the metaphor *no room to swing a cat.* In the days of sail, letting the cat out of the bag (a reference to its canvas carrying bag) was the sign of impending punishment. Its contemporary meaning ashore, that of an untimely revelation, can also bring about undesirable consequences, although it is hoped they are not as serious as a flogging. The use of the brutal cat-o'-nine-tails was outlawed by the United States Congress in 1850 and by the Royal Navy in 1879.

Offing . . . (near in time or space)

Defined by Captain John Smith in his 1627 *Sea Grammar* as "to the seaward or just beyond the anchoring grounds," the word *offing* is used both at sea and ashore to refer to something near at hand or likely to happen in the near future. In her book *Love in a Cold Climate,* Nancy Mitford reflected on "that look of concentration that comes over a French face when a meal is in the offing."

On an Even Keel . . . (smooth sailing)

A ship is said to be in proper trim when she floats in the water on an even keel, that is, neither down by the head nor by the stern in relation to her fore and aft line. A person is said to be "on an even keel" when steady and as well balanced in life as the well-trimmed ship is in the water. Figuratively, a person can "keel over," that is, turn wrong side up, from surprise or shock.

On the Rocks . . . *(impending crack-up)*

Early navigational charts depict rocks as the symbol of danger and destruction. The colloquial expression is an allusion to danger. A marriage on the rocks is one that is battered and foundering on the hardships of life.

Over a Barrel . . . *(in a near-helpless situation)*

There are two possible nautical explanations for this metaphor that describes a situation in which one person is at the mercy of another. Before the development of modern resuscitation techniques, a near-drowning victim was draped face down over a barrel, which was then rolled back and forth in an attempt to revive the victim by draining the water from the lungs. The victim's survival was literally in the hands or control of the rescuer. A second possible nautical source for the contemporary use of the expression stems from the tradition of harsh discipline at sea. Until the mid-nineteenth century, flogging was still an accepted form of punishment in most navies and aboard merchant vessels. William Falconer described how the unfortunate seaman was restrained over the "gunner's daughter," as the barrel of a cannon was called when it was used for this purpose. The sailor was helpless and at the mercy of the jaunty who wielded the lash.

Overwhelm . . . *(overcome)*

From the Middle English word *whelven,* which means to turn upside down, a vessel is said to be "overwhelmed" when she has capsized or has turned upside down in the water. A person is said to be "overwhelmed" when completely overcome mentally or emotionally. Daniel Defoe described this helpless plight in *Moll Flanders*, published in 1721: "I was overwhelmed with the sense of my condition."

Part Brass Rags . . . (a falling-out between friends)

In the days of great wooden sailing ships, sailors shared personal cleaning gear with special shipboard buddies. If they had a falling out, not an unusual event within the confines of a hot, cramped ship, they no longer shared the rags and other gear necessary to polish brass and metal trim. They had literally *parted brass rags.*

Peacoat or Jacket . . . (mariner's traditional jacket)

The short, double-breasted, navy blue, wool or kersey jacket now popular with sailors and landlubbers alike has a long nautical tradition. Its derivation is most likely from the Dutch *pijjekker* (*p* being the sound of the first syllable), a sailor's jacket dating to the fifteenth century, and very similar to the one worn today. The jacket is also known as a "reefer," because its length makes it convenient to wear aloft while reefing or taking in sails.

Pigeonhole . . . (compartmental storage)

Some authors have cited a nautical origin for the figurative pigeonhole, which is a small, open compartment for keeping documents in a desk or cabinet. While it is true that pigeons found the small openings in the ship's capstan to be a comfortable nest, they could be found roosting in any convenient shipboard recess long before the capstan's invention. In 1577, Herrsbach wrote in his work *Husbandry,* "to feed and fat them [doves] in little dark rooms like pigeon holes."

Pipe Down! . . . (be quiet!)

The boatswain's pipe, a distinctively shaped whistle, was once used to transmit commands throughout a ship. Its high-pitched sound was audible above the howling of the wind to crewmen working high in the rigging. Each command had its own particular cadence by which it was identified, understood, and carried out. The cadence to "pipe down" was the last call at night aboard a naval vessel. The signal commanded all unnecessary noise and activity to stop, and hands so assigned to turn in until it was time for them to stand night watch. The ship then became very quiet. The verbal command "Pipe down!" was used by sailors and landlubbers alike when they wanted someone to make less noise or to stop talking. Since the advent of the loudspeaker, the bos'n's pipe is no longer used to issue commands throughout a ship but is limited to the ceremonial piping aboard of visiting VIPs.

Both *pipe* and *cadence* became popular ashore as they strutted and fretted their hour upon the stage (to paraphrase the Bard) in the Rogers and Hammerstein hit *The Sound of Music.* To the horror of Maria, Captain von Trapp's seven children responded to piped cadences—all except little Gretl, who, in spite of being able to hit a B-flat, just couldn't remember her signal.

Plain Sailing . . . (uncomplicated, straightforward)

In the sixteenth century, sailing charts were, according to *The Oxford Companion to Ships and The Sea,* "drawn on the assumption that the earth was flat, even though all experienced sailors knew that it was not." These charts were referred to as "plane," a word derived from the Latin *planum* (a flat surface). Plane (also spelled "plain" until the eighteenth century) sailing charts did not employ the principles of parallels of latitude and meridians of longitude at right angles, required few calculations, and were, therefore, comparatively easy to use. Ashore, "plain sailing" became synonymous with easy progress over a straight, unobstructed course. In 1889, British novelist H. Rider Haggard used the term to describe an easygoing, uncomplicated character: "Healthy, happy, plain sailing Bessie."

Pooped . . . (tired or overwhelmed)

The word *pooped* derives from the Latin *puppis,* meaning the stern or aftermost part of a vessel. Strictly speaking, the "poop" is the name given to the short, aftermost deck raised above the quarterdeck of a sailing ship but is sometimes incorrectly used to describe the nonraised aftermost deck as well. A ship is said to be "pooped" when a wave breaks over her stern as she is running before a gale. "High o'er the poop audacious seas aspire," wrote Scottish poet-lexicographer and sailor William Falconer in his narrative poem "The Shipwreck." In his most notable work, *Universal Dictionary of the Marine,* he described "pooping": "The shock of a high, heavy sea upon the stern or quarter of a ship when she scuds before the wind in a tempest. This circumstance is extremely dangerous to the vessel which is thereby exposed to the risk of having her whole stern beat inwards, by which she would be immediately laid open to the entrance of the sea, and of course, founder or be torn to pieces." The nautical term has washed into the English vernacular in a big way. The word *pooped* is used to describe a person who has taken a figurative beating and is overwhelmed by exhaustion. Ironically, William Falconer drowned in 1769 when his ship, the frigate *Aurora,* foundered off Cape Town, South Africa.

Posh . . . (classy and elegant)

It is widely believed that this much-used acronym originated in the days when British officials, civil servants, and their families traveled regularly to and from India on Peninsular and Oriental Steam Navigation Company ships. The tickets of wealthy and discriminating passengers were stamped with the letters POSH, which meant "port out, starboard home," thereby ensuring accommodations on the cooler side of the ship both ways of the round trip. However, there is no documented evidence to support that tickets ever were stamped in this fashion. It has been suggested that the popular acronym is a corruption of POSN, the initials of the Peninsular and Oriental Steam Navigation Company. The expression was apparently first recorded in print in the 25 September 1918 issue of *Punch:* "Oh yes, Mater. We had a posh time of it down there."

Quarantine . . . (fixed period of enforced isolation of persons, ships, or goods exposed to, or infected with, contagious disease)

During the Middle Ages, contagion ran rampant in port cities. Spread largely by rats from ships, the Black Death, or bubonic plague, wiped out nearly half the population of Europe. Quarantine laws originated in Venice when a Council of Health was assembled to deal with the escalating problem of contagion. The word *quarantine* derived from the Italian *quaranta* (forty)— the number of days that an infected ship was originally held offshore. It is not known how or why the council arrived at that particular number. Since many members of the Venetian Council were learned churchmen, the forty-day waiting period may reflect the traditional period of cleansing as set forth in the Scriptures.

Rats from a Sinking Ship . . . (deserters)

Sailors believed that rats had a sixth sense and that the sight of them leaving a ship in large numbers was a portent of disaster. Even Shakespeare noted the tendency in *The Tempest:*

> *. . . they prepared*
> *A rotten carcass of a boat, not rigg'd,*
> *Nor tackle, sail, nor mast; the very rats*
> *Instinctively have quit it.*

There could have been something to this belief, because rats were denizens of the bilge, the nethermost region inside the ship's hull and the first place to flood when a vessel takes on water. In 1625, Francis Bacon, giving the metaphor a land base, wrote of the "wisdom of rats that will be sure to leave a house somewhat before it fall." When abandoning a cause that seems doomed to failure, a person is considered to be a rat fleeing a figuratively sinking ship.

Real McCoy . . . (authentic, bona fide, and of the highest quality)

Some authorities believe that the expression "real McCoy" originated with an 1890s Chicago prizefighter by the name of Kid McCoy. As the story goes, a saloon heckler questioned the local celebrity's identity. With one mighty blow, the kid decked the heckler who, on regaining his senses, stated: "That's the real McCoy, alright!" Still another version of the expression's origin stems from an Irish ballad dating from around the same period in which the wife of a certain McCoy proclaimed that she was the head of the household, wore the pants in the family, and was, therefore, "the real McCoy." Some authorities link the origin of the expression to a popular post-Prohibition Scotch whisky called McCoy's.

The most widely accepted version of its origin also dates from the days of Prohibition and has a nautical flavor. Bill McCoy, a boatbuilder from the Canadian Maritime Provinces, became very wealthy, not to mention popular, through the smuggling of bootleg liquor to cities along the northeastern seaboard of the United States. As an entrepreneur of some principle, McCoy had no known ties to the organized crime syndicates that flourished during Prohibition. Even more remarkable was the fact that the liquor McCoy delivered was pure, unadulterated, and of the highest quality— an impressive testimonial during a time when the consumption of "home-made hooch" was a frequent cause of blindness and death. Eventually rounded up and convicted of smuggling, McCoy may have been persona non grata to the Feds, but his name became a household word synonymous with 100 percent authenticity and high quality.

Rig . . . (style of dress)

Derived from the Middle English word *rig* (to bind or wrap), the *rig* of a ship denotes the masts, spars, attendant stays or rigging, and the sails that drive the ship. The particular arrangement of the masts and sails differentiates types of vessels—ship, bark, brig, schooner, sloop, and the like—regardless of hull design. The word washed ashore where it is frequently used in reference to costume or style of dress. In his work *A Stray Yankee in Texas,* published in 1853, Paxton mused, "Here was a rig for a July day in Texas with the thermometer at 105 degrees in the shade. . . ."

Rostrum . . . (a speaker's stand)

The bows of Mediterranean warships were once fitted with ramming devices known as *rostra* (beaks). Around 340 B.C., Romans began a tradition of taking *rostra* from enemy vessels and carrying them back home. Highly valued as evidence of Rome's power, the captured *rostra* were displayed in front of the speakers' platform in the Roman Forum. In time, the entire speakers' platform became known as the *rostrum.* The following comment appeared in the 24 May 1976 issue of *The New Yorker:* "Reagan looks good at the rostrum: a tall figure with ruddy cheeks, his reddish brown hair swept back in a slight pompadour." One can only wonder whether the Caesars had such good press!

Round-Robin . . . (a type of elimination system in competition)

Ringleader . . . (instigator)

Round-robin is a corruption of the French words *rond* (round) and *rouban* (ribbon). The term originated in seventeenth-century France, where officials devised a method of signing grievance petitions on ribbons that encircled the documents. The practice was soon adopted by English sailors, who found it the perfect ploy in challenging unnecessarily harsh shipboard conditions and authority. Although there is usually safety in numbers, no one wanted to sign at the top of a formal grievance petition for fear of being held accountable as the chief instigator. By using the *round-robin*—a petition in which names were written in a circle—no name stood out at the top of the list; hence, no one could be identified as the *ringleader*. The custom became so widespread that it was noted in the June 1828 issue of *The Lancet,* England's foremost medical journal: "if thirteen physicians . . . had written what seamen call a round robin to an authority . . ." The round-robin of today describes a tournament in which competitors play against each other at least once. Losing a match doesn't result in immediate elimination, thereby allowing competitors to go around again.

Rub Salt in the Wound . . . (another insult)

In the days of sail, salt was critical to preserve meat stores needed for a voyage that could last many months. It was also used as an antiseptic. An unfortunate sailor who had just received a lashing with the infamous cat-o'-nine-tails (see also "No Room to Swing a Cat") would then have salt rubbed into his wounds. Metaphorically, one rubs salt into a figurative wound by adding insult to injury.

Rummage Sale . . . (sale of cast-off goods)

In the days of wooden merchant ships, dockside warehouses held special sales of damaged cargo. Called *rummage sales,* they took their name from the Old French word *arrumage,* meaning to pack or stow cargo aboard ship. *Rummage* gradually took on the meaning of unwanted, damaged, or low-quality castoffs, such as items of clothing or household goods. The act of "rummaging" also came to mean ransacking or searching through a jumble of objects. Perhaps Charlotte Brontë had a yard sale in mind when she wrote in *Jane Eyre,* "They had been conducting a rummaging scrutiny of the room upstairs."

Running the Gauntlet . . . (assailed from all sides)

Running the gattloppe (narrow path) was a military punishment first recorded in Sweden early in the seventeenth century. The offender was made to strip to the waist and run between facing ranks of comrades who would strike him with belts, whips, or the flat of their swords as he passed. Native Americans were known to inflict a similar punishment on their captives with war clubs. Since the English were familiar with the French word *gauntlet,* meaning a kind of heavy glove, corruption of the Swedish *gattloppe* was inevitable. The expression *running the gauntlet* or *running the glove* makes no sense when compared to running a narrow path. Nevertheless, this expression was used when the practice of running the gauntlet was adopted by the Royal Navy in 1661 as punishment for more serious offenses committed aboard war vessels. In a figurative sense, the expression *running the gauntlet* describes a situation in which one is assailed or criticized from all sides. It was ripe for metaphorical adaptation, and in 1768 Madame D'Arblay commented in her *Early Diary,* "O, what a gauntlet for any woman of delicacy to run."

Sailing under False Colors . . . (to misrepresent)

The word *colors* is used to denote the national flag flown by a ship at sea. Ancient tradition and the law of the sea require that all ships fly their true colors so they can be positively identified. Yet in the days of sail, many a friendly or neutral flag was hauled down moments before an attack and quickly replaced by the ship's true colors. In a colloquial sense, a person sailing under false colors misrepresents or pretends something that, in fact, is inconsistent with the truth. This practice, alas, seems as common on land as it once was at sea.

Scraping the Bottom of the Barrel . . . (low quality)

Life at sea was fraught with hardship. Not only did sailors of old wage a constant war with the elements to survive, but they had to contend with miserable living conditions aboard ship. Not the least of these hardships was their daily food ration, which consisted largely of hardtack biscuit, pork, beef, and fish. Refrigeration and vacuum packing were many years in the future, so a ship's meat supply was preserved in salt or brine and stored in wooden barrels. Since nothing was ever allowed to go to waste during the long months at sea, the ship's cook scraped the fat from the sides and bottom of the barrel when the meat was gone. Small amounts of the congealed fat were used as a lubricant and preservative for leather gear kept aboard ship. The rest of the gelatinous residue was doled out as part of the crew's daily food ration— and the term *scraping the bottom of the barrel* became synonymous with inferior quality. Although the leap from the era of wooden ships to the glamour of Hollywood seems a broad one, the following comment appeared in the 25 October 1993 issue of *People* magazine: "She has gorgeous looks . . . big bucks, fame. So why is it that Julia Roberts always seems to be scratching the bottom of the apparel barrel?"

Scuppered . . . (nonplussed, at a loss)

From the Dutch *schoepen* (to draw off), a *scupper* is a drain hole cut into the bulwarks of a ship to allow water to run off the deck and over the sides. In 1590 the eminent Elizabethan chronicler Richard Hakluyt wrote of "every scupper hole and other place where the rain ran down." A sailor swept off his feet by a large wave breaking over the deck was said to have been "scuppered."

Figuratively, the sliding sailor was about to go down the drain hole with the sea- or rainwater. In his 1896 work *Seven Seas,* Rudyard Kipling metaphorically described the feeling of being at a loss:

We preach in advance of the Army,
We skirmish ahead of the Church,
With never a gunboat to help us,
We were scuppered and left in a lurch.

Scuttle . . . (to abandon or destroy)

Probably from the Old French *escoutilles,* the anglicized word *scuttle* refers to a small hole cut in a hatch cover or in the side of a ship to admit light and air. What is commonly known as a "porthole" to a landlubber is correctly called a "scuttle" by an old salt. *To scuttle a ship* means to sink her deliberately by opening her seacocks (a kind of valve) or by blowing holes in the bottom of her hull so that she fills with water. In June 1919, the most famous incident of mass scuttling occurred at Scapa Flow in the Orkney Islands when the entire German fleet scuttled itself to avoid turning their ships over to the British. In colloquial usage ashore, *scuttling* describes the abandonment or destruction of something. On CNN morning news, 24 February 1994, John Defterios reported that a thirty-million-dollar merger between Bell Atlantic and TCI had been "scuttled."

To *scuttle* has also taken on the meaning of running with short, hurried steps. This kind of hasty retreat was noted by Horace Walpole in a personal letter written in 1739: "We scuttled upstairs in great confusion."

Scuttlebutt . . . (rumor)

During the days of sail, drinking water was kept in a butt or cask in an easily accessible spot aboard ship. In order to ensure that the fresh water would last until the next landing, a "scuttle" or small hole was sawed out of its side so that the butt could only be half-filled at any given time. The forerunner of the office water cooler, the *scuttlebutt,* as it came to be called, was one of the few places aboard a working ship where the sailors could congregate for a moment, relax, and exchange a bit of gossip. U.S. Navy slang for "rumor" or "gossip," the word *scuttlebutt* washed ashore in the 1930s and has been popular with landlubbers ever since.

Sea Change . . . (a change)

By land or by sea, a marked transformation, particularly into something finer, is known as a *sea change.* The expression was immortalized by William Shakespeare in *The Tempest:*

> *Full fathom five thy father lies;*
> *Of his bones are coral made;*
> *Those are pearls that were his eyes:*
> *Nothing of him that doth fade*
> *But doth suffer a sea-change*
> *Into something rich and strange.*

Sea Lawyer . . . (an argumentative sailor)

Definitely not a member of the American Bar Association, a sea lawyer was defined by Admiral William Smyth in *The Sailor's Wordbook* (1867) as "an idle, litigious long-shorer, more given to questioning orders than to obeying them. One of the pests of the navy as well as mercantile marine." Richard Henry Dana referred to them as "lawyers of the foc'sl."

Sea Legs . . . (ability to function with ease)

A sailor is said to have found his *sea legs* when he is able to walk upright, comfortably, and without seasickness on the deck of a rolling ship. In the rollicking Victorian satire on the Royal Navy, Sir William Schwenk Gilbert tells us that the Captain of H.M.S. *Pinafore* was never . . . well, hardly ever, without his sea legs:

> *I am the Captain of the* Pinafore,
> *And a right good captain too!*
> *And I'm never ever sick at sea!*
> *What, never?*
> *No, never!*
> *What, never?*
> *Hardly, ever!*
> *He's hardly ever sick at sea!*
> *Then give three cheers and one cheer more,*
> *For the hardy Captain of the* Pinafore!

A person finds his sea legs ashore when he begins to function in a new or unfamiliar situation with ease and comfort.

Shake a Leg! . . . (hurry up!)

Crews on British warships of old were often denied shore leave when they were in harbor for fear that they would desert. In order to cushion this disappointment, wives and sweethearts were allowed to live on board while the ship remained in port. In the morning, the boatswain's mate would awaken the crew with the cry "Show a leg" (or the variant "Shake a leg"). If the leg that appeared over the edge of the hammock was hairless or had on a stocking, its owner was presumably a woman. The women were permitted to remain in the hammocks while the hairy-leg owners, presumably the sailors, were mustered on deck. The practice of women living aboard British warships was abolished around 1840, but the expression "shake a leg" took on a life of its own ashore, where it still means "Come on, hurry up!"

Shanghai . . . (coerce)

The word *shanghai* was first used by American sailors to describe being knocked unconscious, kidnapped, and impressed into service aboard a ship in need of a crew. Many a sailor who fell victim to the notorious press (impressment) gangs ended up in Shanghai, a major port in the era of sail. The practice was widespread, and an article in a March 1871 issue of *The New York Tribune* noted, "and before that time, they would have been drugged, shanghaied and taken away from all means of making a complaint." In a figurative sense, to be *shanghaied* implies being coerced into a situation against one's will.

Ship . . . (an object that can be navigated)

From the Middle English word *schip* (boat), the word is applied in a metaphorical sense to various objects that can be navigated through an obstacle course to a place of safety, strength, or accomplishment. During the bloody conflict that saw the Union pitted against the Confederate States of America, New Englander Henry Wadsworth Longfellow penned the lines:

Sail on, sail on, O Ship of State!
Sail on, O Union, strong and great!
Humanity, with all its fears,
With all the hopes of future years,
Is hanging breathless on thy fate!

Shipshape . . . (neat and in good order)

The original expression is thought to be *shipshape and Bristol fashion,* dating back to the time when Bristol, England, had the reputation of being the best-organized and most efficient port on

the west coast of England. In his *Seaman's Dictionary* of 1644, Sir Henry Mainwaring wrote, "it being no use for the ship, but only for to make her shipshapen, as they call it."

Ships That Pass in the Night . . . (passing in darkness and silence)

The quintessence of this phrase is contained in *Tales of a Wayside Inn* by Henry Wadsworth Longfellow:

> *Ships that pass in the night*
> *and speak each other in passing,*
> *Only a signal shown*
> *and a distant voice in the darkness;*
>
> *So on the ocean of life,*
> *we pass and speak one another,*
> *Only a look and a voice,*
> *then darkness again and a silence.*

Shiver My Timbers! . . . (expletive denoting surprise or disbelief)

Presumably, this expression alludes to a ship's striking a rock or shoal so hard that her timbers shiver. The expression was first seen in 1834 in the novel *Jacob Faithful* by Frederick Marryat. In 1881, Robert Louis Stevenson found it to be the perfect exclamation for the irascible Long John Silver: "So! Shiver me timbers, here's Jim Hawkins!" This stereotypical pirate expletive became extremely popular with writers of sea yarns and Hollywood swashbucklers.

Shove Off . . . (to go away or leave)

Shoving off is the procedure of moving a ship away from the dock and getting under way. In its land-based colloquial application, *shove off* can be a none-too-polite suggestion roughly equivalent to "go take a powder" or "get lost!"

Sing Out . . . (to articulate words briskly)

Old nautical dictionaries define *singing out* as the chant by which the sailor on the bow of a ship proclaimed his depth soundings with each cast of the lead and line. The expression was illustrated by Admiral Smyth in *The Sailor's Wordbook* of 1867 with the rhyme:

> *To heave the lead the seaman sprung*
> *And to the pilot clearly sung —*
> *By the deep, nine.*

Colloquially, *sing out* is synonymous with the invitation to express an opinion.

Skipper . . . (leader, boss)

The captain or master of a ship is called the *skipper*. The word was introduced in Britain during the fourteenth century and was derived from the Dutch *schipper* (captain), itself derived from *schip,* meaning "ship." The term *skipper* is frequently applied ashore to a person in a position of leadership or authority.

Skylark . . . (frolic)

A "skysail" was the light-weather sail set on a square-rigged ship when the wind was steady and favorable. It was the uppermost sail, and on a tall ship it was called a *skysail* with good reason. Groups of young sailors would play a game of follow the leader, climbing aloft to the skysails then sliding down the backstay to the deck far below. The modern word *skylark,* meaning to indulge in mindless frolic, has its origins both in the skysail and the Anglo-Saxon word *lac,* meaning "to play." In his work *Bracebridge Hall,* published in 1822, Washington Irving described "listening to the lady amateur skylark it up and down through the finest bravura of Rossini and Mozart."

Skyscraper . . . (tall building)

After a third of the city was destroyed by fire in 1871, a new Chicago emerged, phoenixlike, from its own ashes. Erected in 1884, the multistoried, steel-framed Home Insurance Building was an architectural wonder— and the nation's first "modern" structure was promptly dubbed a "skyscraper" by journalists of the day. A hundred years earlier, however, there had been a skyscraper at sea. Used to describe a triangular sail high above squaresails on a ship, "skyscraper" was adapted for use ashore to describe anyone or anything appearing tall enough to touch the sky. In 1826, a sporting magazine reported that riders were seen trotting down a road on "great nine-hand skyscrapers."

Slewed . . . (spinning around, intoxicated)

First recorded as a nautical word, *slue* is described by William Falconer in his 1789 *Dictionary of the Marine* as "to turn any cylindrical or conical piece of timber about its own axis without removing it." The twisting, spinning movement of the nautical procedure has given rise to yet another colloquialism for an intoxicated state. G. Cupples wrote in his adventure *Green Hand,* published in 1849, "we'll save our grog and get slewed as soon as may be."

Sloppy . . . (disheveled)

The word has its origins in the Middle English *sloppe,* meaning a loose-fitting garment. Chaucer used the word to describe a kind of baggy breeches. In a manuscript account of the wardrobe of Queen Elizabeth I, there is an order to John Fortescue for the delivery of some Naples fustian for "sloppe for Jack Greene, our Foole." In 1623, the Royal Navy took note of the fact that many of their crews were in tatters and rags. The Admiralty ordered that ready-made clothing, or "slops," be carried aboard ships and issued to their crews. The baggy, one-size-fits-all garments were kept in a trunk or a locker called a *slop chest.*

After being packed in a damp chest for months on end, the garments that emerged could, indeed, be rather sloppy. The word *sloppy* is now used to describe anything or anyone who is careless, untidy, or slovenly.

◆ ◆ ◆

Slush Fund . . . (general fund for small luxuries)

During the age of sail, fried salt pork was a staple food aboard ship. At the end of a
voyage, the residual grease, or *slush,* was sold in port to candle and soap
makers. Profits were put in the "slush fund," a general account used to
purchase little extras for the crew. The term *slush fund* became very popular
in the political arena in the aftermath of the Civil War when it was first used
to describe a contingency fund set aside by Congress. Since it was outside
their regular operating budget, the "slush fund" was used for highly irregular
and corrupt procedures such as bribes. The expression *pork barrel,* used to
describe government funds appropriated for projects as rewards to loyal
constituents, is probably an offshoot of the post–Civil War "slush fund." In
any case, pork barrel appropriations are frequently contested, having
assumed some of the negative overtones once associated with slush funds.
However, "slush funds" has now reverted to its original meaning, a general
fund for small luxuries.

Snub . . . (to ignore)

From the Middle Scandinavian word *snubba* (a short-stemmed pipe), the word *snub*
describes the process of bringing a line or ship to a halt. A line that is
being payed out freely is usually snubbed by hardening down on a few
loose turns around the bitt. A ship is snubbed by dropping the anchor.
In 1830, sailor-novelist Frederick Marryat used the word (as we still do)
to describe a short, upturned nose: "As my father's nose is aquiline and
mine is a snub . . ." The word is used ashore to describe the act of cutting
someone short by ignoring or treating with disdain. Samuel Richardson
noted such social slights in his novel *Clarissa,* published in 1748: "I must
have been accustomed to snubs and rebuffs from the affluent."

Son of a Gun . . . *(an expletive or term of affectionate regard)*

In the early nineteenth century, wives of sailors in the Royal Navy were permitted to live aboard while the ship was in port. Occasionally they were permitted to accompany their husbands to sea. Because working spaces had to be kept clear, the only place where a woman could give birth aboard ship was behind a canvas shelter jury-rigged between cannons on the gun deck. Admiral William Smyth made a valiant attempt to explain the origins of the expression in *The Sailor's Wordbook*, published in 1867:

> *An epithet conveying contempt in a slight degree, and originally applied to boys born afloat, when women were permitted to accompany their husbands to sea; one admiral declared he literally was cradled under the breast of a gun carriage.*

The expression frequently carried the implication of dubious paternity, explaining the slight degree of contempt to which Smyth referred. In *Customs and Traditions of the Royal Navy*, A. B. Campbell noted a log entry made in 1835 by a British officer commanding a brig off the coast of Spain:

> *This day the surgeon informed me that a woman aboard had been laboring in childbirth for twelve hours and if I could see my way to permit the firing of a broadside to leeward, nature would be assisted by the shock. I complied with the request and she was delivered of a fine male child.*

Such births gave rise to the famous nautical ditty:

> *Begotten in a galley and born under a gun,*
> *every hair a rope yarn, every tooth a marlinspike,*
> *every finger a fishhook,*
> *And his blood right good Stockholm Tar!*

The expression *son of a gun* is still with us, long after the practice of women living aboard ship was abolished by the British Admiralty around 1840.

Son of a Sea Cook . . . (a term of not-so-affectionate regard)

One source gives the following explanation for the origin of the expression *son of a sea cook:* According to author-sailor Robert Hendrickson, the native Algonquian word for skunk was *seganku.* It was appropriated by early English colonists who pronounced the word *segonk.* The Anglicized word also proved to be a tongue twister for the hardy pioneers, and it wasn't long before "son of a segonk" became "son of a sea cook." Although the epithet is still applied to first-rate stinkers, this colorful explanation of its origins probably belongs in the "it's not the truth but it's well invented" category. It is more than likely that *son of a sea cook* is a corruption of "seacock," meaning a "bold sailor" or "sea rover." Captain Frederick Marryat used the expression in *Jacob Faithful,* published in 1835: "Silence you sea cook! How dare you shove in your penny whistle!" (The nautical equivalent of shoving in your two cents worth.)

Sound Out . . . (to seek information)

From the Old English or Old Norwegian word *sund-gyrd,* meaning a sounding pole or line, *to sound* is the process of determining sea depth in the vicinity of a ship. When a whale dives deeply, it is said to be "sounding." The depth of an issue or a person may be sounded out by cautious, indirect questioning. The metaphor was used by Washington Irving in his *Tales of a Traveler,* published in 1824: "He soon sounded the depth of my character."

Splice the Main Brace . . . *(to indulge in happy hour)*

Splicing is a method of joining two ropes or two parts of the same rope together by first unraveling the strands at two ends then joining them by interlocking, then interweaving and tucking the strands. This procedure is an acceptable temporary repair job, but the rope is never considered quite so trustworthy as before the splice. On old sailing ships, the main brace was a line or rope critical to ship handling and subjected to extremes of tension. "Splicing the main brace" was rarely considered and the line would be replaced, rather than spliced, when it was worn. "Splicing the main brace" was a traditional and figurative expression in the Royal Navy for serving an extra tot of rum on rare, special occasions. The expression was another cynical analogy, for like the literal splicing of the main brace, its figurative counterpart was just as rare an event. Rum rations were discontinued in the Royal Navy in 1970, but the expression and the metaphorical activity live on in full vigor ashore.

Staunch . . . *(strong, standing firm to principle)*

From the Old French *estanche* (holding tight or impervious to water), the metaphorical use of the word was applied by William Shakespeare in Act II of *Antony and Cleopatra:*

> *Yet if I knew*
> *What hoop [hope] should hold us stanch, from edge to edge*
> *O' the world I would pursue it [truth].*

Stem to Stern . . . *(from one end to the other)*

The stem is the foremost timber or steel member forming the bow of a vessel; the stern is the after end of the vessel. In *Idylls of the King*, Tennyson described Arthur's sombre barge as being "dark as a funeral scarf from stem to stern." When used ashore in the figurative sense, the expression *stem to stern* means thoroughness—a search of a place from one end to the other.

Stinkpot . . . *(foul, noxious)*

From the Old English *stincan,* meaning to disperse or emit, a *stinkpot* was a chemical warfare bomb of great antiquity. Use of the stinkpot was universally popular with buccaneers who ranged the seven seas. A clay pot was charged with a concoction of chemicals, and a slow-burning fuse was inserted. The fuse was ignited and the stinkpot tossed or dropped onto the enemy ship. Sailor and lexicographer William Falconer described how the suffocating fumes "caused turmoil and distraction as the attacking ship came alongside with its boarding party." As an insult, "stinkpot" was used by early-nineteenth-century critic J. King in an article entitled "The Beauties of the Edinburgh Review, alias, the Stink-pot of Literature."

Stranded . . . (in an unfavorable position)

From the Old English *strand* (land bordering on the sea) and the Dutch *stranden* (to drive ashore or run aground), one is said to be "stranded" when forced into an unfavorable situation without means of coping adequately. In 1851, writer-artist-critic John Ruskin commented on the work of seventeenth-century French landscape artists Claude Lorrain and Nicholas Poussin: "the works may be left, without grave indignation, to their poor mission of furnishing drawing-rooms and assisting stranded conversations."

Sun Is over the Yardarm . . . (time for happy hour to begin)

This expression is thought to have had its origins in an officers' custom aboard ships sailing in the North Atlantic. In those latitudes, the sun would rise above the upper yards—the horizontal spars mounted on the masts, from which squaresails were hung—around 11 A.M. Since this coincided with the forenoon "stand easy," officers would take advantage of the break to go below for their first tot of spirits for the day. The expression washed ashore where the sun appears over the figurative yardarm a bit later in the day, generally after 5 P.M., and the end of the workday.

Swab . . . (mop up)

From the Middle Dutch *swabbe* (mop), the word *swabber* was used in the late sixteenth century in reference to sailors who cleaned the decks. William Shakespeare used the term in a song in *The Tempest*:

> *The master, the swabber, the boatswain and I,*
> *The gunner and his mate,*
> *Loved Mall, Meg and Marian and Margery,*
> *But none of us cared for Kate . . .*

As a reference to a merchant seaman, the word *swabbie* was seen in print as early as 1835 in Frederick Marryat's *Jacob Faithful*. *Swabbie* came into general usage during the mid-nineteenth century after it was used by Herman Melville in his novel *White Jacket*. Both "swab" and "swabbie" have become popular slang terms for sailors, particularly those of the U.S. Navy.

Swallow the Anchor . . . (to give up life at sea and settle down on land)

This is exactly what Shakespeare's character Stephano sings about in *The Tempest*: "I shall no more to sea, to sea, / Here shall I die ashore."

Swamped . . . (overwhelmed)

First recorded by Captain John Smith in 1624, *swamp* was a term peculiar to the North American colony of Virginia. The word probably had been in local use in England prior to that time, where it was used to describe low-lying ground where water collected. A ship is said to be "swamped" when she is filled with water and in danger of sinking.

Figuratively, an individual is "swamped" when significantly outnumbered or overwhelmed and unable to complete the tasks at hand as scheduled or desired. The horseless Richard III, facing the superior forces of Henry, Earl of Richmond, at Bosworth field, was clearly swamped. His inability to cope against insurmountable odds resulted in his demise and the introduction of the Tudor dynasty.

Tack . . . (a maneuver; to maneuver)

On the Right or Wrong Tack . . .
(having made a correct or incorrect maneuver)

In nautical parlance, the word *tack* has many meanings, only one of which is used figuratively by sailors and nonsailors alike: to progress from point A to point B by making a series of moves that reflect a stratagem. A vessel sailing against the wind travels forward by making a series of zigzag movements. That process, called *tacking,* requires changing the position of the sails to alter the direction of the ship. Without skillful maneuvering, the vessel might stay on one tack (zig) longer than necessary and pass the critical point at which another change in tack (zag) would have brought her closer to her goal. Once the critical point has been passed, the ship is said to be on the "wrong tack." In his poem "Christmas at Sea," Robert Louis Stevenson described a ship leaving home port and tacking for the open sea:

We gave the South a wider berth, for there the tide-race roar'd;
But every tack we made brought the North Head close aboard;
So's we saw the cliffs and houses, and the breakers running high,
And the coastguard in his garden, with his glass against his eye.

Metaphorically, a person is on the wrong tack when taking the wrong approach to a situation or an issue. To correct the error and take the right tack, the course of action must be altered. An abrupt reversal of policy usually casts doubt on the credibility or skill of the policy maker. As a consequence, a certain amount of hedging or metaphorical tacking is usually employed— the eulogy in Shakespeare's *Julius Caesar* a case in point. When Mark Antony stood over the corpse of Caesar and addressed a public supportive of the emperor's murder, he tacked his way through the eulogy and used the point of his verbal zigs and zags to turn the mob against the murders.

Taken Aback . . . (surprised by circumstances)

By definition, a square-rigged vessel is said to be "taken aback" when her sails billow out in reverse due to a sudden, unforeseen wind shift or inattention on the part of the helmsman. The vessel's forward motion stops and everything is reversed. In a figurative sense, one is "taken aback" by a sudden and surprising turn of events.

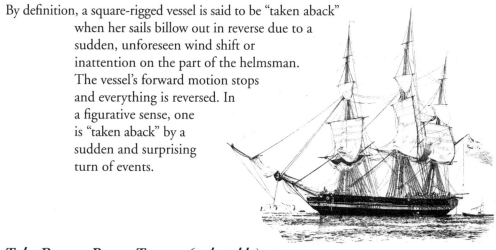

Take Down a Peg or Two . . . (to humble)

This well-worn metaphor is thought to have its origins in post-Armada England. In the aftermath of the stunning victory in 1588 over the reputedly invincible Spanish Armada, England's naval pride, as well as her pride in personal achievement, soared. During this period flags and pennants began to play an all-important role in indicating the official rank and personal status of the ship's commander. Flags were hoisted on halyards (small ropes or lines) and secured to one of a series of small pegs arranged vertically on the mast. The higher the flag was flown from the mast, the higher the honor. When a commander handed over the ship to a subordinate officer, or one of inferior personal status, the new commander's flag would be flown lower down the mast. According to tradition, the flag had to be *taken down a peg or two.* It was noted in the court and times of Charles I (1625): "the Bishop of Chester that bore himself so high, should be hoisted a peg higher."

Take the Wind Out of Her Sails . . . (to slow down or bring to a standstill)

When a sailing ship is deprived of the wind's driving power, she slows down or stops dead in the water. This standstill can also occur from the deliberate maneuver of another ship. As one ship passes to the windward of another, she "blankets" the wind blowing on the other's sails, thus depriving her of driving power. This tactic was used in warfare during the days of the great sailing navies and is still common today in competitive sailing. Taking the figurative wind out of a person's sails remains a very popular land-based metaphor, which means to curtail another's forward momentum in order to outdistance the competition.

Tar . . . (a term of affectionate regard for a sailor)

First recorded in the mid-seventeenth century, the expression originates from the seafarer's custom of treating his clothing with tar as a protection against the elements. In a popular eighteenth-century ballad, a lass whose mother wanted her to marry a wealthy landowner sings,

*I know you'd have me wed a farmer
And not give me my heart's delight.
Mine's the lad whose tarry trousers,
Shine to me like diamonds bright.*

Teetotaler . . . (total abstainer from alcoholic drink)

This well-known expression is an emphasizing, reduplication, or extension of the word *total.* In a June 1840 publication of the *American Temperance Union,* Dr. W. Patton commented, "total abstinence from all intoxicating drinks is a principle of English manufacture so they adopted what they call a teetotal pledge." According to Admiral William Smyth, the word *teetotaler* may have nautical origins. In *The Sailor's Wordbook,* published in 1867, Smyth defined "teetotaler" as "a very old amplification of totally, recently borrowed from sea diction to mark a class who totally abstain from alcoholic drink."

Tell It to the Marines . . . (expression of disbelief)

Dating from the mid-seventeenth century, this expression has been attributed by William Falconer, and other early sources, to King Charles II of England. Apparently the king had great confidence in the knowledge and skill, as well as the veracity, of his newly formed "Marine Regiment of Foot," and is credited with saying, "henceforth, ere we cast doubts upon a tale that lacks likelihood, we will first tell it to the marines." As the words passed from the court of the monarch to the decks of his ships, the expression was given a derogatory connotation by ever-cynical sailors who considered marines to be gullible fools. *Tell it to the marines* was recorded in this context by Lord Byron in his narrative poem *The Island* (1823): "that kind of talk will do for the marines but sailors are much too bright to believe it."

As an expression of disbelief, *tell it to the marines* became so firmly entrenched in English-speaking vernacular that newspaper correspondent Malcolm Muggeridge used it to send a cryptic message to the Western world. During the height of the Cold War, the Moscow-based Muggeridge wrote an article on Russia's defense budget that, due to Kremlin censorship, was largely fiction. Much to the delight and enlightenment of American and British readers, Muggeridge concluded with, "suggest you inform the marines." Perhaps the most unforgettable use of the expression came from Anna Magnani in her Academy Award–winning performance in the film version of the Tennessee Williams play *The Rose Tattoo*. In the role of widow Serafina Delle Rose, Magnani waxed incredulous at the antics of the dim-witted but lovable hero, Alvaro. Because she spoke no English, Magnani had memorized her lines phonetically—and "tell it to the marines" was given another dimension by the great Italian actress when she hissed at the grinning Burt Lancaster: "Heh! Go tail heet to da muddrrreeenzza!"

Three Sheets to the Wind . . . (an intoxicated state)

On a square-rigged sailing ship, a "sheet" is a line attached to the lower corners of a squaresail, used for trimming it to the wind. When sheets are allowed to run free, the sails lose their wind and flap and flutter. The ship's forward motion stops, and as she loses steerageway, she becomes impossible to control. A person is said to be "three sheets to the wind" when in an advanced state of inebriation, fluttering and wallowing around out of control. "Shaking a cloth in the wind" and "being over the bay" are nautical expressions for being only slightly drunk.

Tide Over . . . (to bide time)

To tide over, a method of working the tides in the English Channel, was first described by Captain John Smith in 1627 in *A Sea Grammar.* Proceeding down channel, square-rigged ships of the period could make little progress against the incoming tide and prevailing southwesterly wind. In order to work the tidal currents to their best advantage, ships anchored during the flood (east-running tide), weighed anchor at high water, then beat to windward on the strength of the ebb (west-running tide). Later, William Falconer defined the term in his 1789 *Dictionary of the Marine* as "alternately sailing and anchoring, depending on the tide, in order to work a ship in or out of port." To "tide over" and its variant, "tide it out during a lull," mean to wait for the next opportunity to make progress on the flowing tide. The expression was seen in print as early as 1659 and was still current in 1821, as the Earl of Dudley noted in a letter: "I wish we may be able to tide over this difficulty."

Time and Tide Waiteth for No Man . . . (seize the day)

The word *tide* has its origins in the Old English *tid*, meaning "time" or "an extent of time." Nautically speaking, *tide* is defined as the alternate rise and fall of the earth's oceans on a coast. Tides are caused by the gravitational attraction of the sun and moon. The expression *tidal current* describes the inflow and outflow of ocean waters that usually accompany the rise or fall of the tide. Colloquially, *tide* can refer to both the vertical and horizontal movement of water. In the following lines from *Julius Caesar,* Shakespeare's Brutus admonishes that time and tide waiteth for no man and that success lies in taking "the current when it serves"—in going with the flow:

There is a tide in the affairs of men,
Which, taken at the flood, leads on to fortune;
Omitted, all the voyage of their life
Is bound in shallows and in miseries.
On such a full sea are we now afloat,
And we must take the current when it serves,
Or lose our ventures.

Brutus was speaking of the strategies of war, but his advice is just as sound when applied to the eddies and currents, the ebbs and flows, of life's circumstances. In common usage as early as the twelfth century, the expression "time and tide waiteth for no man" is basically an alliteration of two more or less synonymous words.

Touch and Go . . . (an "iffy" situation)

According to early nautical dictionaries, a ship that makes a very short stay in a port while en route to her final destination is said to "touch and go." In a general sense, the expression describes the act of doing something for a minute, then quitting it immediately. When about to engage an enemy, Lord Nelson employed his own maxim: "Touch and take!" Referring to alcoholic tonics, in 1655 Moufet and Bennett admonished that "howsoever we may taste of it to bring our appetite, let it be but touch and go." The expression *touch and go* is today most commonly applied to a highly uncertain or precarious situation or condition, in which the slightest turn will result in failure or disaster, or as Admiral Smyth noted, "within an ace of ruin."

Touch It with a 10-Foot Pole . . . (to keep at a distance)

The origin of the expression "I wouldn't touch it with a 10-foot pole" is uncertain. It is thought to derive from the 10-foot-long barge poles that river boatmen used to push their craft along in shallow water. While a subordinate use of the poles is to fend off things, there is no real evidence to support this theory.

Trick . . . (taking one's turn at something)

As early as 1669, a *trick* was defined in Sturmey's *Mariner Magazine* as the time allotted to a man on duty at the helm of a ship. In colloquial usage, "trick" refers to a prostitute's customers. The analogy inherent in its use is relatively straightforward.

True Blue . . . (honest and loyal)

This popular expression was defined by Admiral William Smyth in *The Sailor's Wordbook* as "a metaphorical term for an honest, hearty sailor true to his uniform, and uniformly true."

Turn a Blind Eye . . . *(to ignore intentionally)*

During the Battle of Copenhagen in 1801, Admiral Horatio Nelson deliberately placed
his telescope to his blind eye so that he could not see an order to break off
action with the enemy. Realizing that to withdraw would mean failure in an
important campaign, Nelson ignored the order and engaged the enemy.
The result left England with a stunning victory and posterity with a
"dandy" metaphor. The redoubtable Nelson remarked later that he had
a blind eye so he sometimes had the right to use it.

Under the Weather . . . *(a case of the "blahs")*

Many expressions adapted for colloquial usage ashore have grown out of mariners'
eternal battle with the elements. *Under the weather bow* was a reference to
the side of the ship's bow that was taking the full brunt of head seas in foul
weather. It was also an expression applied by sailors to seasickness.

Walk the Chalk . . . *(follow the rules)*

The ability to walk a straight chalk line drawn along the deck of a ship was the standard
sobriety test in many navies. *Walking the chalk* carries the same connotation
ashore but has taken on the additional meaning of a strict adherence to
rules.

Wallop . . . *(a sound thrashing)*

During the reign of King Henry VIII (1509–1547), a French fleet raided and burned
the town of Brighton on the coast of Sussex, England. A furious Henry
ordered the soldier-diplomat Sir John Wallop to carry out a reprisal raid
on the coast of Normandy. Wallop did so with enormous efficiency, and
demolished many French towns, villages, and harbors. It has been said that
Wallop's name became synonymous with "a thorough beating" because of
this action. It is a wonderful story but, alas, most likely not true. The word
wallop derives from the fourteenth-century Middle English word *wallope*,

which means a horse's gallop, a violent heavy movement, or a heavy, resounding blow. The colorful word must have been a source of amusement for poet Charles E. Carryl when he wrote "The Walloping Window-Blind":

A capital ship for an ocean trip
Was the "Walloping Window-blind" —
No gale that blew dismayed her crew
Or troubled the captain's mind.

Waster or Waister . . . (a loafer)

The nautical derivation of *waster* or *waister* to mean "a loafer" stems from a combination of two words: *waist* and *waste.* The Old English *waest,* originally a reference to the shape of fruit, especially the narrow end of a pear, evolved to include the size and shape of the middle of the human body. Edmund Spenser wrote in 1579, "gird your waste, with a tawdry lace." Sailors expanded the meaning to include the middle of a ship. According to Young's *Nautical Dictionary* (1846), "waisters" were "greenhands or broken down seamen placed in the waist [middle section] of a ship of war to do duty not requiring a knowledge of seamanship." In his novel *White Jacket,* Herman Melville referred to "waisters" as the "tag-rag and bobtail of the crew" who "haul aft the fore and main sheets, besides being subject to ignoble duties; attending to drainage, etc." The word *waster,* a derivative of the Old French *wastere* and Old English *guaste* (one who ravages or dissipates), clearly describes a less-than-competent individual. Such usage was in print as early as 1350. No doubt the fact that English spelling was not standardized until the seventeenth century played a significant role in creating the two versions of the word.

Weather Eye . . . (keen observation)

Weathering a Storm . . . (survive in good condition)

The weather or windward side of a ship takes the brunt of head seas in foul weather. A "weather eye" was explained by Admiral William Smyth in his *Sailor's Wordbook* as "the eye which is specially used for observing the weather." In his *Storm at Hastings* (1839), Thomas Hood observed that "his weather eye the seaman aimed across the calm, and hinted by his speech a gale the next morn."

Used in figurative speech, such as "to keep a weather eye opened," the expression means "to be on guard," or "on the alert" for change. H. Rider Haggard did not have literal meteorological forecasting on his mind when he wrote in *She* (1887), "Job returned in a great state of nervousness, and keeping his weather eye fixed on every woman who came near him." Many colloquial expressions adapted for general use have arisen from sailors' descriptions of raging elements at sea. When weathering a figurative storm, for example, a person comes safely through a period of trial and tribulation. In 1626, John Donne noted such metaphorical activity: "That soul which is but near destruction, may weather that mischief [storm]."

Windfall . . . (good luck in the form of an unexpected acquisition)

In medieval times, trees could not be cut down in England on tracts of land specified as timber reserves for shipbuilding, a critical national interest. However, if a tree was felled by the wind, the owner of the land could use the timber for his own purposes.

Wishy-Washy . . . (inconstant)

The phrase *wishy-washy* describes a person who is weak and indecisive. The usage stems from old nautical dictionaries that define *wishy-washy* as any drink that is too weak. Note one such entry from Admiral William Smyth's 1867 *Sailor's Wordbook:* "His food, the land crab, lizard or frog / His drink a wish-wash of six water grog."

WIND, WAVES, AND WEATHER

When people first took to the sea, they began to amass knowledge and develop the skills necessary for survival in a dynamic, frequently hostile environment. Not the least of these survival skills was the ability to predict the weather. Changes in wind direction, the appearance of the sky, cloud characteristics, and other naturally occurring phenomena all had special significance for the experienced mariner. Over the centuries, prognosticating skills were expressed by anonymous sailors in hundreds of easy-to-remember jingles and rhymes.

Here is just a small sampling:

Sea-gull, sea-gull sit on the sand!
It's never good weather when you're on the land.

Trace the sky with painter's brush,
The winds around you will soon rush.

Sound traveling far and wide,
A stormy day will betide.

At sea with low or falling glass [barometer],
Soundly sleeps the careless ass,
Only when its high and risin',
Safely rests the careful wise 'un.

This body of knowledge based on practical experience was widely known and highly respected by the general populace. A cartoon in the *World Almanac* of 1878 suggests that when newfangled technical methods for predicting weather fail, old lore remains reliable. The cartoon depicts a frog sitting in a large glass jar that is labeled "weather frognosticator," an allusion to the belief that frogs always croak loudly before a soaking rain.

Henry Wadsworth Longfellow took a more serious approach in his poem "The Wreck of the Hesperus." The poem's title refers to the results of the captain's folly in ignoring the old sailor's advice.

Then up spake an old sailor
Who had sailed the Spanish Main,
I pray thee put into yonder port,
For I fear a hurricane.

Last night the moon had a golden ring,
And tonight we no moon see!
The skipper, he blew a whiff from his pipe,
And a scornful laugh laughed he.

The most universally known of all the weather jingles was adapted from chapter 16 of the Gospel According to St. Matthew. When Jesus is asked for a sign from heaven, He responds: "When it is evening ye say, it will be fair weather for the sky is red. And in the morning it will be foul weather today for the sky is red and lowering." This wisdom survives, paraphrased for sailors and landlubbers alike, as:

Red sky in the morning,
Sailor take warning,
Red sky at night,
Sailor's delight.

"Blowing great guns and small arms" is an old nautical saying that refers to gale- or hurricane-force winds. At sea, such winds build up mountainous waves, which crash with a hollow booming noise. Figuratively speaking, "blowing great guns" gets the idea across even if you don't hear the noise.

In *The Sailor's Wordbook*, Admiral William Smyth defined a devil's smile as "a gleam of sunshine amongst dark clouds either in the heavens or in the captain's face."

"Fog dogs" describe transient prismatic breaks in thick mists and clouds, considered by sailors to be good signs for clearing weather.

Sailors had a low opinion and scant regard for all landlubbers, and soldiers in particular. Their contempt is implied in the expression *soldier's wind,* a breeze that blows on the beam of a vessel under sail, providing such favorable conditions that tacking or trimming the sails is not necessary and no particular nautical skill is needed in sailing the ship. The expression was seen in print for the first time in *Peter Simple* (1833) by Captain Frederick Marryat: "The wind is what is called at sea soldier's wind, that is blowing so that ships could lie either way, so as to run out or into the harbor." As is the case with many of the salty expressions used by Marryat, it is not known whether the sailor-novelist invented "soldier's wind," or simply repeated something that was already part of the sailor's vernacular.

There is an ancient belief that waves occur in a pattern, becoming progressively higher and stronger, culminating in the ninth (or seventh) wave, after which the progression begins again. The classic expression of this belief is found in *The Idylls of the King,* in which Alfred, Lord Tennyson described the coming of King Arthur on a thunderous ninth wave:

> . . . *And then the two*
> *Dropt into the cove, and watch'd the great sea fall,*
> *Wave after wave, each mightier than the last,*
> *Till last, a ninth one, gathering half the deep*
> *And full of voices, slowly rose and plunged*
> *Roaring, and all the wave was in a flame;*
> *And down the wave and in the flame was borne*
> *A naked babe, and rode to Merlin's feet,*
> *Who stoopt and caught the babe, and cried,*
> *"The King!" . . .*

From the Arthurian legend and Queen Victoria's poet laureate to pop music artists, the big wave rolls on. According to Sting, the seventh wave is the ultimate—it is love!

YARNS OF THE SEA, LEGENDS, MYTHS, AND SUPERSTITIONS

For all the romance associated with it, life at sea was once fraught with hardship and danger. As superstitious European sailors ventured north into vast expanses of uncharted waters, they were beset by doubt and fear of the unknown. Sailors believed that the more they learned about the mysteries of the sea, the better their chances of survival. To that end, practical information was passed from sailor to sailor, from one generation to another. And as mariners began ranging the seven seas, they collected beliefs and traditions, frequently adapted or embellished the tales in accordance with their own beliefs and customs, and passed on their new-found information to others. Like the yarn fibers in a well-made rope, tales were spun and interwoven into an incredibly rich body of sea lore. Characters in Frederick Marryat's *Jacob Faithful* beg an old salt for a story with the words "Come spin us a good yarn, Father." Indeed, good yarns abound.

The practice of christening a ship by breaking a bottle of champagne over her bow has its origins in pagan ritual. Ancient Norsemen and Romans believed that in order to ensure good fortune for the vessel and her crew, the keels of their warships had to "taste" the blood of live persons during launching. Slaves or captured prisoners were tied to the keel blocks and, as the ship slipped down the launching ramp into the water, it crushed the unfortunate victim under its weight. As centuries passed and society became slightly more squeamish, red wine became a symbolic substitute for blood, and so developed the custom of pouring wine on the deck and bow of a ship as an offering to the gods of the sea. According to a cherished but undocumented tradition, one of the most famous ships in our history, the U.S. frigate *Constitution,* steadfastly refused to budge when teetotaling New Englanders tried to christen her with plain water. It was only after a bottle of fine old Madeira was produced and whacked across her bow that "Old Ironsides" slid gracefully down the ways into Boston Harbor.

Women were considered to bring bad luck at sea and it was widely believed that the sea gods grew angry at the very sight of them on board ships. However, this superstition did not preclude the wives of British seamen from once living aboard ships with their husbands, and occasionally, in the merchant service, wives accompanied their ship captain husbands to sea. Paradoxically, mariners believed that a naked woman before the ship had the power to calm gales and high winds. For this reason, many a ship's figureheads depicted a woman with one bare breast. Although women were considered to bring bad luck, mariners always use the pronoun "she" when referring to their ships. Whether its proper name is masculine, or whether it is a man o'war, a battleship, or a nuclear submarine, a ship is always referred to as "she." This very old tradition is thought to stem from the fact that in the Romance languages, the word for "ship" is always in the feminine. For this reason, Mediterranean sailors always referred to their ship as "she," and the practice was adopted over the centuries by their English-speaking counterparts. One source suggests that a ship "was nearer and dearer to the sailor than anyone except his mother." What better reason to call his ship "she"?

A "cat's paw" describes a ruffle on the water during a calm that moves as silently as a cat. In *Jacob Faithful,* Frederick Marryat described how "cats paws of wind, as they call them, flew across the water here and there, ruining its smooth surface." On seeing cats' paws on the water, old salts would rub the ship's backstay, part of the ship's standing rigging, as though stroking a cat, and whistle for a wind to come to the ship. But to whistle on board when the wind was blowing would bring bad luck, for it was believed that whistling mocked the devil and that he would retaliate by sending gale-force winds. It was also widely believed that, because the priest was the devil's natural enemy, Satan would attack a ship bearing a priest by sending violent winds.

Sailors believed that possession of the caul of a newborn baby was a sure prevention against death by drowning. In Chapter One of Charles Dickens's *David Copperfield* (1850), the hero relates, "I was born with a caul which was advertised for sale in the newspaper at the low price of 15 guineas." Such advertisements appeared in newspapers and journals well into the twentieth century.

The phenomenon known as St. Elmo's fire is an electrical discharge that takes place around the mastheads and yardarms of a ship under certain atmospheric conditions. According to Italian legend, the fourteenth-century bishop, martyr, and patron saint of Mediterranean sailors, St. Elmo (whose name is thought to be a corruption of the name "Erasmus"), was rescued from drowning by a sailor. As a token of his gratitude, St. Elmo promised to send a light to warn those at sea of approaching storms. Mediterranean seafarers of the fifteenth and sixteenth centuries believed that the ghostly light emanated from the body of Christ. Called the *Corposanto*, or "Holy Body," by the Italians and Portuguese, the word was corrupted to "corposants" by English-speaking sailors and was immortalized in the following passage by Herman Melville in *Moby Dick*:

> *"Look aloft!" cried Starbuck, "The St. Elmo's Lights,*
> *Corposants! The corposants!" All the yardarms were tipped*
> *with a pallid fire; and touched at each tri-pointed lightning-rod*
> *end with three tapering white flames, each of three tall masts*
> *was silently burning in that sulphurous air, like three gigantic*
> *wax tapers before an altar.*

For the most part, St. Elmo's fire was considered a favorable omen, but many older mariners believed that if the eerie, shimmering light fell upon a man's face, he would die within twenty-four hours.

The tall masts of ships are excellent conductors of lightning, and superstitious mariners of old used many charms to protect themselves against its dangers. After St. Barbara's cruel father beheaded her with his own hands for her Christian beliefs, he was struck dead by lightning. The name of "St. Barbara" is still invoked today against the perils of lightning and fire.

In many cultures, it was widely believed that the dead return to help or harm the living. Legends of spectral or ghost ships were nautical manifestations of this fearful belief—and one of the most famous is that of the Flying Dutchman. A Dutch captain by the name of Vanderdecken (variant: Van Dyck) swore by "Donner" and "Blitzen" ("thunder" and "lightning") that he would put in to Table Bay in spite of a raging gale that he attributed to the wrath of God. Vanderdecken's ship foundered while the terrible oath was still on his lips. As a result of his blasphemy and defiance of Divine wrath, the Dutchman and his crew were condemned to wander the seas for eternity. A ballad published in 1848 described the plight of the unfortunate captain:

> *Moan, ye Flying Dutchman, for horrible is thy doom:*
> *The ocean round the stormy Cape is thy living tomb;*
> *For there Van Dyck must beat about forever, night and day:*
> *He tries in vain his oath to keep to anchor in Table Bay.*

Over the years, sightings of the Flying Dutchman were reported by many levelheaded, courageous sailors who believed that anyone who laid eyes on the spectral ship would die within twenty-four hours or be struck blind.

A variation on the Flying Dutchman legend tells of a Captain Faulkenberg, condemned to sail the North Sea forever, playing dice with the devil (or death) for his soul. Sir Walter Scott used the theme of Faulkenberg's plight in his narrative poem *Rokeby*, as did Samuel Taylor Coleridge in *The Rime of the Ancient Mariner*:

> *Her lips were red, her looks were free,*
> *Her locks yellow as gold:*
> *Her skin was white as leprosy,*
> *The Night-mare Life-in-Death was she,*
> *Who thicks man's blood with cold.*
> *The naked hulk alongside came,*
> *And the 'twain were casting dice;*
> *"The game is done! I've won! I've won!"*
> *Quoth she and whistles twice.*

The eternal wanderings of the Dutchman and his phantom ship have such wide appeal that they have been the inspiration for innumerable works, including sea ballads, novels, short stories, poems, and a grand opera. In 1843, the German composer Richard Wagner wrote an opera based on the legend, but he gave the story a new twist. Forever preoccupied with the theme of redemption, Wagner's story allows the doomed Dutchman to come ashore once every seven years to search for a woman whose love and faithfulness alone would be his salvation. Vanderdecken finds his true love in a Norwegian fishing village a long way from Table Bay and the Cape of Good Hope. A 1920s Metropolitan Opera production of Wagner's *Flying Dutchman* was reported to have ended in grand operatic style with a sinking, storm-tossed ship—St. Elmo's fire shimmering from her mastheads—blood-red sails billowing out beneath a howling, unearthly wind, and the frenzied crew chanting *"Hoe! Huissa!"* (roughly the German equivalent of "heave-ho!"). As the curtain fell on the chaos of Wagnerian symbolism, the redeemed captain and his bride were seen in a last embrace, not going down with the ship as dictated by tradition, but soaring high above it, "transfixed by love."

Traditional sea lore abounds with stories of fantastic half-human, half-beast water creatures. Sirens were creations of classical mythology, part bird and part woman, who perched on rocks or islands. According to mythology, they enchanted mariners with melodious songs and thus lured them to the source of the music—hazardous rocks where, of course, the ships sank and the sailors perished. In other instances, sailors were so captivated by the sound of the sirens' voices that they stopped all their work to listen. Unable to sail their ships away from the mesmerizing sound, the sailors eventually died of starvation. The Greek hero Odysseus outwitted the sirens by instructing his crew to seal their ears with wax so that they would be unable to hear the deadly singing. Never one to miss out on any adventure, Odysseus did not seal his own ears but had himself lashed to the mast so he could not interfere with the orderly running of the ship as she passed within earshot of the enchanting sound. The sirens were so outraged at being foiled in their attempt to seduce Odysseus and his men that they flung themselves into the sea and drowned. Odysseus outwitted the sirens at Cape Pelorus in Sicily, which was subsequently renamed Cape Sirenis after the enticing ladies of the perilous rocks.

When sirens immigrated to the New World, they headed for southern California. Because of unearthly wailing and eerie songs heard as they sailed past it, fishermen and sailors of the gold rush days thought Santa Barbara Island was inhabited by sirens. The first settlers in the region discovered, however, that the enchanting melodies were coming not from the sirens but from the island's sole inhabitants—cats! It seems that some seagoing "mousers" had been marooned when their ship foundered offshore. The "siren song" was, in fact, the commotion made by their progeny as they squabbled over dead fish and other tidbits of kitty cuisine. In modern society, a "siren" still signals danger, but the siren itself has been reformed and gone respectable—it warns rather than lures. The signal device, invented in 1819, was called a "siren" because of the way its sound carried across water.

Mermaids and their male counterparts, mermen, are mythological denizens of the deep with a history dating to antiquity. Legends of creatures with the head and trunk of a man or woman and the lower torso of a fish are an integral part of the folklore of virtually every country with a seafaring tradition. Early writers frequently confused the mermaid with the siren of classical mythology. In 1481, William Caxton wrote, "they be called Sirens or Mermaidens." Shakespeare perpetuated the mistake in *The Comedy of Errors*:

◆ ◆ ◆

O, train me not, sweet mermaid, with thy note,
To drown me in thy sister's flood of tears:
Sing, siren, for thyself and I will dote:
Spread o'er the silver waves thy golden hairs.

Mermen, often considered to be the spirits of sailors lost at sea, were depicted as ugly old men with straggly black beards and hair. No wonder the bulk of mer-lore centered on mermaids, who were believed to be young, attractive, and playful creatures with long golden hair. According to legend, mermaids long for an immortal soul but can attain their goal only through physical union with a human. It is not surprising, therefore, that most mermaid lore centers around the theme of seduction— mermaids trying to entice sailors to live with them at the bottom of the sea. It is said that Captain John Smith (of Jamestown and Pocahontas fame) had a narrow escape when he saw a mermaid in the Caribbean. After long months at sea, the captain was bewitched by the sight of a beautiful woman swimming with sensuous grace alongside his ship. Just as Smith was about to "go overboard"—so to speak—the mermaid flipped out of the water, revealing her scaly lower body and fishlike tail. The legend does not elaborate on the reaction of the hot-tempered Captain Smith.

Mermaid folklore seems to have some basis in reality, springing from the existence of actual aquatic mammals. The manatee and the dugong often take on a touching human appearance, particularly when nursing their young. They were often mistaken for mermaids by lonely, homesick sailors. In his poem "Sailor Man," H. Sewell Bailey mused on the mental isolation of a sailor at sea:

Wayward as a seagull,
Lonely as a hawk
Yet he believed in fairies
And heard the mermaids talk.

Legend gives many forms to the water creature known as a "kelpie," but the name itself is thought to stem from the Gaelic *calpach* (steer or colt). The kelpie is said to appear on land as a beautifully saddled and bridled horse who charms unwary travelers, particularly children, into mounting him. The kelpie then returns to the depths of the sea, where the rider can live but never again return to land. To women, the kelpie appears as a handsome young man with seaweed in his hair. But the kelpie is most frequently depicted as a black horse with red eyes who, according to popular legend, rises from the murky depths to warn of impending disasters at sea. Thought to inhabit every lake and stream in Scotland and Wales, as well as the Eastern Shore of Maryland on this side of the Atlantic, some local folklore depicts the kelpie as a malevolent equine spirit who lures unsuspecting travelers to their doom.

The folklore of Scotland's Hebrides and Orkney Islands, as well as that of Norway, abounds with tales of "silkies," enchanted seal-man creatures, noted for their power to foretell the future. It is said that the silkies live in the depths of the sea but come up on land where, after divesting themselves of their sealskins, they pass as mortal men. Some families living in the Hebrides and Orkneys claim that they can trace their ancestry to a human manifestation of a silkie. According to a haunting, traditional ballad, "The Great Silkie of Sule Skerrie," a seal-man fathered a son by a mortal woman, and then foretold his own death and that of his child:

> *I am a man upon the land,*
> *And I am a silkie in the sea;*
> *And when I'm far and far from land,*
> *My dwelling is in Sule Skerrie.*
>
> *And it shall come to pass on a summer's day,*
> *When the sun shines hot on every stone,*
> *That I will take my little young son,*
> *And teach him for to swim the foam.*
>
> *And thou shall marry a gunner proud,*
> *And a proud gunner I'm sure he'll be,*
> *And the very first shot that e're he shoots,*
> *He'll shoot both my young son and me.*

In Greek mythology, a nymph was a minor divinity who lived on land or in the sea. They were not immortal but extremely long-lived and were of a benevolent nature, kindly disposed toward mortals. Scylla was a beautiful singing sea nymph who, by the machinations of the jealous Circe, was turned into a six-headed dog-monster and rooted in a cave (variant: rock) on the Italian side of the Straits of Messina. Scylla was not pleased with her gruesome makeover and, consumed with bitterness, destroyed everything that came near her. The loss of her beauty did not affect the fascinating voice (or voices) that Scylla used to sing mariners to their doom. Charybdis was not a sea nymph but a greedy mortal woman who stole Hercules' oxen. For her misdeed, Charybdis was turned into a dangerous whirlpool on the Sicilian side of the Straits of Messina, located opposite Scylla's cave. To this day, Scylla and Charybdis form a navigational hazard in the Straits of Messina. Their names have became synonymous with two perilous alternatives, where one cannot be avoided without incurring equally great peril from the other.

A drum said to have been used by Sir Francis Drake during the battles with the Great Armada of 1588 is now at his old estate of Buckland Abbey in Devon, England. According to legend, Drake promised on his deathbed that the drum would beat out a warning if ever England was in danger of an invasion by sea and that he himself would again come to her defense. Drake's Drum was the subject of a famous poem by Sir Henry Newbolt, published in the *Jones Street Gazette* in 1897 and subsequently set to music by Ralph Vaughn Williams:

> *Take my drum to England, hang 'et by the shore,*
> *Strike 'et when your powder's runnin' low;*
> *If the Dons sight Devon, I'll quit the port o'heaven,*
> *An' drum them up the Channel as we*
> *Drummed them long ago.*

It is said that the ghostly beating of Drake's Drum was last heard just before the German fleet scuttled itself at Scapa Flow in the Orkney Islands on 21 June 1919.

Fiddler's Green is the sailor's name for the waterfront district in a large seaport, as well as the name of a sailor's traditional afterlife. This secular heaven is considered to be a place of free-flowing rum, tobacco, taverns, dance halls, and similar amusements, all reminiscent of activities sailors enjoy while on shore leave. In his novel *The Dog Fiend or Snarley-yow* (1837), Captain Frederick Marryat wrote:

> *At Fiddler's Green, where seamen true,*
> *When here they've done their duty,*
> *The bowl of grog shall still renew,*
> *And pledge to love and beauty.*

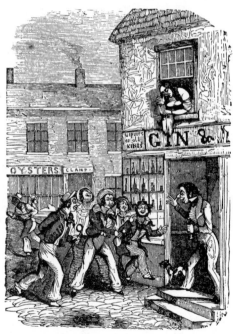

A song written in 1966 by John Conolly, "The Fiddler's Green," is so popular in Great Britain that it is often considered to be traditional by those who are not familiar with its relatively recent origins. This author heard it for the first time in 1981, in Helston, Cornwall, and subsequently in Searsport, Maine— seafaring communities on opposite sides of the Atlantic. Conolly's nautical Camelot, a place "where sailors go if they don't go to hell," is a peaceful haven for shipmates to lie at leisure while the skipper brews their tea.

Fiddler's Green still holds a special mystique for men of seafaring tradition. It continues to buoy their spirits through danger, misfortune, and every other hardship of a rigorous life at sea. And when old salts cut their painters, they drift silently and peacefully away to Fiddler's Green.

BIBLIOGRAPHY

Ashley, Clifford. *The Ashley Book of Knots*. New York: Doubleday, 1944.

Bartlett, John. *A Collection of Passages, Phrases and Proverbs Traced to Their Sources in Ancient and Modern Literature*. 125th ed. Boston, Toronto, and London: Little, Brown and Company, 1980.

Beavis, Bill, and Richard McCloskey. *Salty Dog Talk, The Nautical Origins of Everyday Expressions*. London and New York: Granada Publishing, Adlard Coles, Ltd., 1983.

Campell, A. B. *Customs and Traditions of the Royal Navy*. Aldershot, Hampshire: Wellington Press, Gale-Polden, Ltd., 1956.

Chapman, Robert L., ed. *Thesaurus of American Slang*. Grand Rapids, Mich.: Harper and Row, 1989.

Colcord, Joanna Carver. *Sea Language Comes Ashore*. Cambridge, Md.: Cornell Maritime Press, 1945, 1974.

Craig, Hardin, ed. *Shakespeare*. Rev. ed. Glenview, Ill.: Scott, Foresman and Company, 1958.

Dictionary of National Biography from the Earliest Times to 1900. London: Oxford University Press, 1921–1922.

Evans, Bergan, and Cornelia Evans. *A Dictionary of Contemporary American Usage*. 3rd printing. New York: Random House, 1957.

Falconer, William, A. *Universal Dictionary of The Marine, or A Copious Explanation of the Techniques and Phrases Employed in the Construction, Equipment, Furniture, Machinery, Movements and Military Operation of a Ship.* A new edition corrected. London: T. Cadell, Publishers in the Strand, 1789.

Friedman, Albert B., ed. *The Viking Book of Folk Ballads of the English Speaking World.* New York: Viking Press, 1963.

Green, Jonathon. *Dictionary of Contemporary Slang.* New York: Stein and Day, 1984.

Harry, Lahaina, ed. *Rhyming in the Rigging: Poems of the Sea.* Woodbridge, Conn.: Ox Bow Press, 1978.

Hendrickson, Robert. *Salty Words.* New York: Hearst Marine Books, 1984.

Hourigan, Lt. P. W., U. S. N. *Manual of Seamanship for Officer of the Deck, Ship Under Sail Alone.* 1903 original edition. Annapolis, Md.: Naval Institute Press, 1981.

Kemp, Peter, ed. *The Oxford Companion to Ships and the Sea.* London, New York, and Melbourne: Oxford University Press, 1976.

Mainwaring, G. E., and W. G. Perrin. *The Life and Works of Sir Henry Mainwaring (1587–1653), A Seaman's Dictionary.* London: Navy Records Society, 1922.

Makkai, Adam. *A Dictionary of American Idioms.* 2d ed. Chicago: University of Illinois at Chicago, 1986.

McArthur, Tom, ed. *The Oxford Companion to the English Language.* Oxford and New York: Oxford University Press, 1992.

Murray, Sir James, A. M., ed. *A New English Dictionary on Historical Principles, Founded Mainly on the Material Collected by the Philological Society.* Oxford at Clarendon, 1909.

Onions, C. T., C. W. S. Freidrichson, and R. W. Burchfield, eds. *The Oxford Dictionary of English Etymology.* London: Oxford at the Clarendon Press, 1964.

The Oxford English Dictionary. 20 vols. Oxford at the Clarendon Press, 1989.

Palmer, Roy, ed. *The Oxford Book of Sea Songs.* Oxford: Oxford University Press, 1986.

Partridge, Eric, ed. *Dictionary of Slang and Unconventional English.* 7th ed. New York: The MacMillan Company, 1970.

Rogers, John. *Origins of Sea Terms.* Mystic, Conn.: Mystic Seaport Museum, Inc., 1985.

Smith, Captain John. "A Sea Grammar," in *The Complete Works of Captain John Smith (1580–1631) in Three Volumes.* Edited by Phillip L. Barbour. 3 vols. Chapel Hill and London: University of North Carolina Press, 1986.

Smyth, Admiral William H. *The Sailor's Wordbook, An Alphabetical Digest of Nautical Terms, Including Some More Especially Military and Scientific, But Useful to Seamen; as well as Archaisms of Earlier Voyages.* London: Blackie and Sons in Paternoster Row, 1867.

Webster's Third New International Dictionary. Springfield, Mass.: Merriam Webster, Inc., 1986.

INDEX